RIVER GEOMORPHOLOGY

INTERNATIONAL ASSOCIATION OF GEOMORPHOLOGISTS

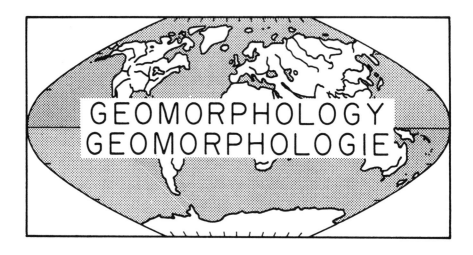

Publication No. 1

THE EVOLUTION OF GEOMORPHOLOGY
A Nation-by-Nation Summary of Development

Edited by H. J. Walker *and* W. E. Grabau

Publication No. 2

RIVER GEOMORPHOLOGY

Edited by Edward J. Hickin

RIVER GEOMORPHOLOGY

Edited by

EDWARD J. HICKIN
Simon Fraser University, Canada

JOHN WILEY & SONS
Chichester · New York · Brisbane · Toronto · Singapore

Copyright © 1995 by John Wiley & Sons Ltd,
Baffins Lane, Chichester,
West Sussex PO19 1UD, England

Telephone National 01243 779777
International (+44) 1243 779777

All rights reserved.

No part of this book may be reproduced by any means,
or transmitted, or translated into a machine language
without the written permission of the publisher.

Other Wiley Editorial Offices

John Wiley & Sons, Inc., 605 Third Avenue,
New York, NY 10158-0012, USA

Jacaranda Wiley Ltd, 33 Park Road, Milton,
Queensland 4064, Australia

John Wiley & Sons (Canada) Ltd, 22 Worcester Road,
Rexdale, Ontario M9W 1L1, Canada

John Wiley & Sons (SEA) Pte Ltd, 37 Jalan Pemimpin #05-04,
Block B, Union Industrial Building, Singapore 2057

Library of Congress Cataloging-in-Publication Data

River geomorphology / edited by Edward J. Hickin
    p. cm. — (Publication / International Association of
  Geomorphologists ; no. 2)
    Includes bibliographical references and index.
    ISBN 0-471-95531-0
    1. Rivers — Congresses. 2. Geomorphology — Congresses. I. Hickin,
Edward J. II. Series: Publication (International Association of
Geomorphologists); no. 2.
GB1201.2.R58 1995
551.3'55—dc20   94-37464
                                                                                                      CIP

British Library Cataloguing in Publication Data

A catalogue record for this book is available from the British Library

ISBN 0-471-95531-0

Typeset in 10/12 pt Times by
Mathematical Composition Setters Ltd, Salisbury, Wiltshire, SP3 4UF
Printed and bound in Great Britain by
Bookcraft (Bath) Ltd

Contents

Contributors		vii
Series Preface		ix
Preface		xi
Chapter 1	Size Characteristics of Sediment Eroded from Agricultural Soil: Dispersed Versus Non-Dispersed, Ultimate Versus Effective **M. C. Slattery and T. P. Burt**	1
Chapter 2	Dune Geometry and Sediment Transport: Fraser River, British Columbia **R. A. Kostaschuk and S. A. Ilersich**	19
Chapter 3	Results of Bedload Tracer Experiments in a Mountain River **K.-H. Schmidt and D. Gintz**	37
Chapter 4	The Interrelations Between Mountain Valley Form and River-Bed Arrangement **C. de Jong and P. Ergenzinger**	55
Chapter 5	Effective Discharge for Bedload Transport in a Subhumid Mediterranean Sandy Gravel-Bed River (Arbúcies, North-East Spain) **R. J. Batalla and M. Sala**	93
Chapter 6	Within-Reach Spatial Patterns of Process and Channel Adjustment **S. N. Lane, K. S. Richards and J. H. Chandler**	105
Chapter 7	Modelling Contemporary Overbank Sedimentation on Floodplains: Some Preliminary Results **A. P. Nicholas and D. E. Walling**	131
Chapter 8	Hydraulic Geometry and Channel Scour, Fraser River, British Columbia, Canada **E. J. Hickin**	155

Chapter 9	Torrential Flow Frequency and Morphological Adjustments of Ephemeral Channels in South-East Spain **C. Conesa García**	169
Chapter 10	Channel Changes on the Po River, Mantova Province, Northern Italy **D. Castaldini and S. Piacente**	193
Chapter 11	The Impact of River Regulation, 1850–1990, on the Channel and Floodplain of the Upper Vistula River, Southern Poland **A. Lajczak**	209
Chapter 12	Morphological Changes in a Large Braided Sand-Bed River **E. Mosselman, M. Huisink, E. Koomen and A. C. Seijmonsbergen**	235
Index		249

Contributors

R. J. Batalla GRAM, Mediterranean Environmental Research Group, Departament de Geografia Física, Universitat de Barcelona, 08028 Barcelona, Catalunya, Spain

T. P. Burt School of Geography, University of Oxford, Mansfield Road, Oxford, OX1 3TB, UK

D. Castaldini Dipartimento di Scienze della Terra, Universitá degli Studi di Pisa, Via S. Maria, 53, 56126 Pisa, Italy

J. H. Chandler Engineering Photogrammetry Unit, Department of Civil Engineering, Loughborough University of Technology, UK

C. Conesa García Department of Geography, University of Murcia, 30001 Murcia, Spain

C. de Jong Berlin Environmental Research Group (BERG), Institut für Geographische Wissenschaften, Freie Universität Berlin, Grunewaldstrasse 35, 12165 Berlin, Germany

P. Ergenenzinger Berlin Environmental Research Group (BERG), Institut für Geographische Wissenschaften, Freie Universität Berlin, Grunewaldstrasse 35, 12165 Berlin, Germany

D. Gintz Geomorphologisches Laboratorium, Freie Universität Berlin, Altensteinstrasse 19, Berlin, Germany

E. J. Hickin Department of Geography and the Institute for Quarternary Research, Simon Fraser University, Burnaby, British Columbia, Canada V5A 1S6

M. Huisink Institute for Earth Sciences, Free University, De Boelelaan 1085, 1018 HV Amsterdam, The Netherlands

S. A. Ilersich Department of Geography, University of Guelph, Guelph, Ontario, Canada, N1G 2W1

E. Koomen Kastelenstraat 243 3h, 1082 EG Amsterdam, The Netherlands

R. A. Kostaschuk Department of Geography, University of Guelph, Guelph, Ontario, Canada, N1G 2W1

A. Lajczak Research Centre for Nature Protection, Polish Academy of Sciences, Lubicz Str. 46, 31-512 Cracow, Poland

S. N. Lane Department of Geography, University of Cambridge, Downing Place, Cambridge, CB2 3EN, UK

E. Mosselman Delft Hydraulics, PO Box 152, 8300 AD Emmeloord, The Netherlands

A. P. Nicholas Department of Geography, University of Exeter, Exeter, EX4 4J, UK

S. Piacente Dipartimento di Science della Terra, Università degli Studi di Modena, Corso Vittorio Emanuele II, 59, 41100 Modena, Italy

K. S. Richards Department of Geography, University of Cambridge, Downing Place, Cambridge, CB2 3EN, UK

M. Sala GRAM, Mediterranean Environmental Research Group, Departament de Geografia Física, Universitat de Barcelona, 08028 Barcelona, Catalunya, Spain

K.-H. Schmidt Geomorphologisches Laboratorium, Freie Universität Berlin, Altensteinstrasse 19, Berlin, Germany

A. C. Seijmondsbergen Landscape and Environmental Research Group, University of Amsterdam, Nieuwe Prinsengracht 130, 1018 VZ, Amsterdam, The Netherlands

M. C. Slattery Department of Geography, East Carolina University, Greenville, North Carolina 27858-4353, USA

D. E. Walling Department of Geography, University of Exeter, Exeter, EX4 4J, UK

Series Preface

The Third International Geomorphology Conference was held at McMaster University, Hamilton, Ontario, in August of 1993. No consistent format has yet been established for publishing papers presented at these conferences. Short contributions from the first conference (Manchester) were published in two large volumes, and selected papers from the second conference (Frankfurt) have appeared in a number of special issues of journals. Although some papers from the third conference will also be published in special editions, independently of the International Association of Geomorphologists, others will appear in a series of volumes dealing with specific geomorphological themes. *River Geomorphology* is the first of the volumes from this meeting to be published by John Wiley & Sons, Ltd in the "International Association of Geomorphologists" series. All the submitted papers in this volume were independently reviewed before being accepted for publication. It is anticipated that four other volumes, dealing with geomorphological hazards, steeplands, the geomorphological effects of minor climatic change, and plenary sessions will also be published as a result of this Conference.

A. S. Trenhaile, Series Editor
University of Windsor, Ontario

Preface

Rivers everywhere have a special place in the study of the earth. Rivers not only represent a vital resource for human activity but they are central to an understanding of the geomorphology of most regions. These important conduits carrying water and eroded materials from the continents to the oceans exert a fundamental control on the shape of the land. Geomorphologists in many countries have long explored these fluvial process/landform relations, sometimes from the perspective of the scientist concerned with understanding the ways of nature, and at other times from the perspective of the engineer concerned with controlling river behaviour. This strong tradition of fluvial studies in geomorphology continues as an important theme in modern earth science research and is clearly reflected in the contributions to the Third International Geomorphology Conference of the International Association of Geomorphologists held in Hamilton, Ontario, Canada, in August 1993.

This volume contains a selection of river studies developed from contributions to the Canadian conference of the International Association of Geomorphologists. They reflect the rich diversity of perspectives and interests of river scientists internationally. The collection opens with a paper by Slattery and Burt (UK) on the textural character of sediments supplied to rivers and is followed by papers by Kostaschuk and Ilersich (Canada), Schmidt and Gintz (Germany) and a companion paper by de Jong and Ergenzinger (Germany) on studies of bedload transport and channel form in rivers. Batalla and Sala (Spain) examine the pattern of effective discharge and bedload transport and Lane, Richards and Chandler (UK) present a model of the process/form linkage within a river channel. A similar process/form connection is sought by Nicholas and Walling (UK) in their study of overbank sedimentation of floodplains. The relation between channel scour and hydraulic geometry is explored by Hickin (Canada) and examples of long-term channel changes, particularly in response to human impacts, are examined by Conesa García (Spain), Castaldini and Piacente (Italy) and Lajczak (Poland). The collection concludes with observations on large-scale channel changes derived from satellite imagery by Mosselman, Huisink, Koomen and Seijmonsbergen (The Netherlands).

I am very grateful to the contributors who made this volume possible. I am also indebted to colleagues who gave so generously of their time in assisting contributors to bring their work to publication. In particular I would like to acknowledge the efforts of Michael Church, Joe Desloges, Wayne Erskine, Ken Gregory, Janet

Hooke, Ian Hutchinson, David Knighton, Michael Lapointe, Dan Moore, Gerald Nanson, Ken Page, Geoff Pickup, Arthur Roberts, Michael Roberts, Ken Rood, Margaret Schmidt, Olav Slaymaker, Derald Smith and Des Walling.

Edward J. Hickin
11 July 1994

1

Size Characteristics of Sediment Eroded from Agricultural Soil: Dispersed Versus Non-Dispersed, Ultimate Versus Effective

M. C. SLATTERY* AND T. P. BURT

School of Geography, University of Oxford, UK

ABSTRACT

This study examines the size characteristics of sediment eroded by concentrated overland flow from valley-side slopes in an agricultural catchment. Samples of surface runoff were collected periodically and analysed for the size distributions of the eroded (undispersed) sediment. This sediment was subsequently dispersed and the size distribution of the primary particles was also determined. Samples of the original surface soil were also dispersed to determine textural size distribution. For all samples, the undispersed sediment was coarser than either the dispersed soil or dispersed sediment, indicating that much of the sediment eroded as aggregates. Medium (63–250 μm) and large (>250 μm) aggregates contained both silt and sand but small amounts of clay. Most of the clay eroded as primary clay which is significant from the standpoint of sediment-bound pollutant transport. The size distribution of eroded material was also related to flow conditions. In some cases, flow in rills was only competent enough to transport the finer silt and clay aggregates at the expense of the coarser sand-sized material. However, most flow was fully competent to transport all the material supplied to it with greater discharges capable of transporting a larger percentage of aggregated sediment. In such cases, the sediment eroded with a "texture" that was quite different to that of the matrix soil. Of significance from this study were the large errors that resulted when the transportability of sediment was inferred from the dispersed textural characteristics rather than the eroded undispersed sediment.

INTRODUCTION

The transportability of sediment by runoff and its potential for subsequent deposition depend largely on its size distribution. Although particle size data are frequently

---

*Current address: Department of Geography, East Carolina University, Greenville, North Carolina 27858-4353, USA

available for soils and sediment, these are commonly evaluated after the sediment is fully dispersed into its primary particles. Such data may be termed *ultimate* particle size data, since they refer to the discrete particles comprising the sediment. However, there is increasing evidence to indicate that most of the sediment moving through a drainage basin will be transported as aggregates rather than as discrete particles. Aggregates are generally much larger than the primary soil particles of which they are composed. Thus, the size distribution of eroded sediment in the field, often termed the *effective* particle size distribution (after Ongley *et al.*, 1981), may be quite different from the size distribution of the dispersed matrix soil or from the size distribution of the primary particles (sand, silt and clay fractions) that result when the eroded sediment is dispersed. It is therefore crucial to consider the effective particle size of sediment, because this will govern the actual behaviour of the transported sediment.

Very few data are available on the size distribution of sediment in the form that it is eroded in the field. Work carried out by scientists at the USDA (e.g. Meyer *et al*, 1980, 1992; Young, 1980; Alberts *et al.*, 1983; Foster *et al.*, 1985) has provided some important information on the size characteristics of eroded sediment from agricultural soils in the United States. However, most of this research has been conducted on small experimental plots under simulated rainfall. There are virtually no data on the characteristics of sediment eroded under natural conditions and there has been no examination of these characteristics over entire hillslopes. A notable exception is the work of Parsons *et al.* (1991) in southern Arizona, although, in this study, sediment was still dispersed in the laboratory because the sandy catchment soils were not well aggregated and tended to erode mostly as primary particles. However, most of the sediment eroded from agricultural soil, as noted above, is composed of aggregated particles. Walling (1990), in a review of future research needs in the field of erosion and sedimentation, stated that the collection of information on the effective size of transported sediment represents an important prerequisite for the development of effective modelling procedures and for improved understanding of the enrichment mechanisms associated with sediment delivery.

The present study was undertaken against this background. The aim of the investigation was to provide more detailed information on the size characteristics of sediment in the form that it is eroded in the field under natural conditions. We also examine the relationship between sediment particle-size characteristics and surface runoff dynamics and calculate the amount of sediment that could be carried by different rates of flow depending on whether the soil is dispersed or in aggregated form.

FIELD SITE AND METHODS

The study catchment (UK grid reference: SP 356 362) is situated in the north-eastern Cotswolds, approximately 8 km north-east of the town of Chipping Norton (Figure 1.1). The catchment has a drainage area of 6.2 km^2 and its altitude ranges from 126 m OD at its outlet to 202 m OD on the northern divide. Slopes are predominantly gentle (around 1°) near the interfluves but steeper (>5°) in the central part of the basin. The stream network is moderately incised into the valley floors,

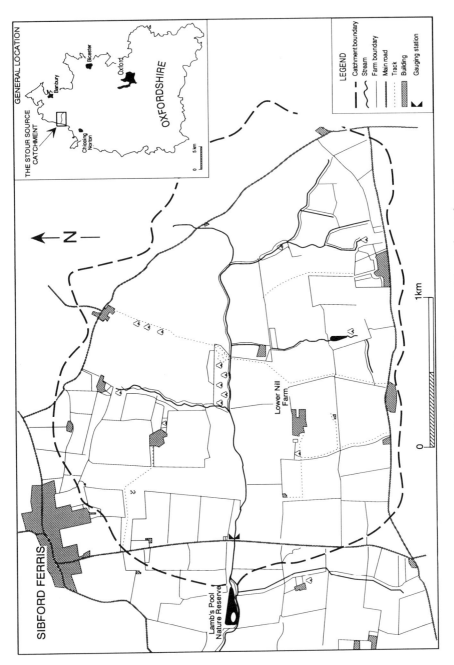

FIGURE 1.1 Map showing location of the study catchment

and bank heights are commonly less than 1 m, although deeper incision occurs in places. The area is underlain by Middle Jurassic sediments, predominantly Northampton Sand and Great Oolite Limestone, which have a regional dip of a few degrees to the south-east (Horton *et al.*, 1987). Two major soil types have been identified within the basin (Jarvis *et al.*, 1984). *Brown calcareous earths* of the Aberford Series occupy the central portion of the basin; these are moderately stony, well-drained fine loamy soils over the limestones at moderate depth. *Ferritic brown earths* of the Banbury Series occur as bands across the northern and southern sections of the basin; these are stony, well drained, fine loamy soils with bright ochreous iron-rich subsoil. Land use is predominantly mixed arable farming.

A gauging station was established at the catchment outlet in July 1992. A 90° V-notch weir with an Ott stage recorder was installed to provide continuous measurements of stream discharge. Two rain gauges were also installed in the field: a tipping-bucket gauge at the catchment outlet (this was linked to a Campbell data logger and provided the time of each 0.2 mm of rainfall) and a Casella autographic natural siphon gauge near Lower Nill Farm (see Figure 1.2). We also installed a Rock and Taylor automatic pump water sampler just upstream from the gauging station. Water and suspended sediment samples were obtained from the stream every 4 hours during low flows and every 15 minutes during storm events.

Catchment slopes were monitored for surface runoff and erosion between July 1992 and June 1993. Slope runoff was generally of two types: (i) concentrated overland flow along vehicle wheel tracks (wheelings), and (ii) flow in rills. Concentrated flow along wheelings was observed frequently during the very wet months of December 1992 and January 1993. These flows were generally thin (less than 1 cm depth) with discharges always less than 1 l s^{-1}. Flow velocities along the wheelings ranged between 0.16 m s^{-1} and 0.25 m s^{-1}. Thus, only small amounts of sediment were transported downslope in such flows. However, three fields along the southern slopes of the basin produced considerable amounts of surface runoff during the winter storms (Figure 1.2). In sub-catchment A on what we termed Field A, closest to the basin divide, downslope wheelings generated runoff on several occasions eventually eroding the soil to produce a braided rill network (Figure 1.3). In the centre of Field A, in sub-catchment B, an extensive thalweg rill system developed. Here, the rill network consisted of a single channel developed along the valley floor (Figure 1.4) which was supplied with runoff and sediment by a series of "feeder rills" developed along wheelings on the steeper valley-side slopes. This rill system extended into the adjacent field downslope (i.e. Field B, Figure 1.2), although here the network contained only a single channel in the valley floor. During a storm on 13 January 1993, flow from the rill system upslope reached Field C, concentrating along a wheeling to form a single channel with a fan deposit directly adjacent to the stream. Thus, the entire rill system, once fully developed, formed a continuous and efficient link for the delivery of water and sediment between the upper fields and the stream channel.

Runoff was sampled at five locations across the three fields shown in Figure 1.2. Two or more samples from each site were averaged to obtain the reported data. Samples were collected by hand by placing a 0.5 l plastic bottle into the flow. The samples were returned to the laboratory immediately following each storm event and

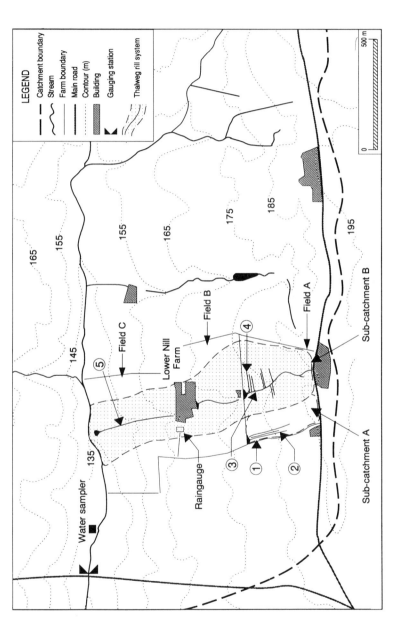

FIGURE 1.2 Sampling sites across the three fields referred to in the text. Site 1 was located along a small rill and Site 2 along a large rill in sub-catchment A on Field A (these rills are also shown in Figure 1.3). Site 3 was located in the main thalweg rill and Site 4 in one of the feeder rills (see Figure 1.4). Site 5 was located in the rill on Field C, adjacent to the stream

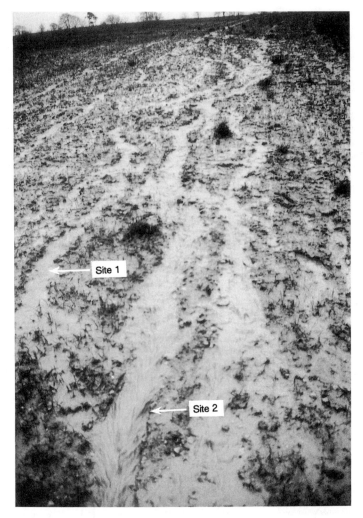

FIGURE 1.3 The braided rill network in sub-catchment A with sampling sites 1 and 2. The photograph was taken during the storm of 13 January 1993

were first analysed for the size distributions of the eroded undispersed sediment. The samples were wet-sieved through a nest of sieves with 2000, 1000, 500, 250, 125 and 63 μm openings to determine the content of sand-sized sediment. Wet-sieving consisted of gentle, thorough sieve-by-sieve washing of sediment, using ample clean water to flood each sieve. The material passing through the 63 μm sieve (i.e. the silt/clay fraction) was then transferred to a large beaker. Five subsamples were taken from the silt/clay fraction by pipette and run through a Cilas 920 laser granulometer to determine the mean, non-dispersed sediment sizes at 31, 16, 8, 4 and 2 μm. The remaining sediment in the beaker was then dried at 50°C and weighed.

After determining the undispersed sediment size distributions, the eroded sediment from each sample was combined into one of three sediment size groups:

FIGURE 1.4 The thalweg rill in sub-catchment B taken during the storm of 13 January 1993

(a) coarser than 250 µm, (b) 63–250 µm, and (c) finer than 63 µm. The sediment in each of these large, medium and fine size groups was subsequently dispersed with sodium hexametaphosphate and then sieved and run through the laser granulometer to obtain the primary particle content of each group. A sample of the top 30 mm of surface soil was also collected near each of the runoff sampling sites. The primary particle size distribution of this soil after dispersing was determined using the techniques described above.

In addition to sampling runoff and sediment, we also made several hydraulic measurements in the rills during storms; the parameters recorded were the flow velocity, water depth and rill width. Flow velocities were measured by dye injection using potassium permanganate as a tracer. Following Abrahams et al. (1986), these

TABLE 1.1 Size distributions of dispersed surface soil, sediment as eroded in runoff, and this sediment after dispersal. Note that the samples are identified as follows: the first number indicates the sample site (see Figure 1.2); the second number identifies the date of sampling (18 for the storm of 18 December 1992 and 13 for the storm of 13 January 1993)

Particle size class (μm)	Dispersed soil	Sample 1.18[a]		Sample 2.18[b]		Sample 2.13[c]		Sample 3.13[d]		Sample 4.13[e]		Sample 5.13[f]	
		Eroded sediment	Dispersed sediment	Eroded sediment	Dispersed sediment	Eroded sediment	Dispersed sediment	Eroded sediment	Dispersed sediment	Eroded sediment	Dispersed sediment	Eroded sediment	Dispersed sediment
1000–2000	0.3	0.0	0.0	0.5	0.0	2.4	0.1	2.5	0.2	2.1	0.0	2.3	0.0
500–1000	0.7	0.8	0.0	1.1	0.3	3.1	0.4	4.5	0.8	4.4	0.2	7.0	1.2
250–500	2.6	1.6	0.0	6.1	2.0	5.9	1.8	8.6	4.0	8.6	1.9	12.8	2.9
125–250	14.9	5.8	1.6	24.9	18.9	29.0	25.7	39.5	33.7	33.3	24.5	15.7	8.1
63–125	19.1	7.0	2.3	5.9	7.5	10.1	9.1	9.7	12.5	13.4	13.9	15.7	9.3
31–63	12.8	14.8	10.5	9.3	8.5	12.4	11.4	7.8	6.5	9.1	6.7	16.4	21.7
16–31	13.8	18.9	19.4	12.2	13.0	12.3	15.2	8.5	11.7	8.3	14.4	9.9	16.3
8–16	12.7	21.5	23.7	15.6	16.6	10.3	14.7	7.9	11.5	8.1	14.5	8.0	15.5
4–8	10.5	15.4	21.2	12.4	15.9	7.2	10.9	5.6	9.4	6.3	12.1	5.9	12.3
2–4	6.7	8.0	12.2	6.7	9.9	4.1	6.1	3.0	5.5	3.7	7.0	3.5	7.2
<2	5.9	6.3	9.0	5.2	7.3	3.2	4.7	2.4	4.2	2.7	4.9	2.9	5.4

[a] Sample taken from small rill at Site 1 (Figure 1.2) on 18/12/92
[b] Sample taken from large rill at Site 2 (Figure 1.2) on 18/12/92
[c] Sample taken from large rill at Site 2 (Figure 1.2) on 13/01/93
[d] Sample taken from thalweg rill at Site 3 (Figure 1.2) on 13/01/93
[e] Sample taken from feeder rill at Site 4 (Figure 1.2) on 13/01/93
[f] Sample taken from rill adjacent to stream at Site 5 (Figure 1.2) on 13/01/93

data were multiplied by 0.7 or 0.8, depending on whether the flow was transitional or turbulent, to give mean velocities. However, during highly turbulent and turbid flow, particularly in the main rill channel, the dye quickly became diluted and we had to resort to timing the passage of corks introduced into the flow. Depth of flow in the deepest section of the rills and rill width were determined by rule measurements (see Abrahams et al., 1986; Abrahams and Parsons, 1990; Slattery and Bryan, 1992 for a more detailed discussion on hydraulic measurements in rill and interrill flows). All the results presented in this paper pertain to sediment transported in runoff in suspension.

ANALYSIS OF SEDIMENT SIZE CHARACTERISTICS

The size distributions of sediment eroded from the catchment slopes in surface runoff, as indicated by sieve and granulometric analysis, are given in Table 1.1 along with the size distribution of the dispersed matrix soil (column 1). The primary particle size distributions for the portion of sediment (a) coarser than 250 µm, (b) between 63–250 µm, and (c) finer than 63 µm are given in Table 1.2. The size distribution of the dispersed sediment shown in Table 1.1 (column 2 for each sample) was obtained by compositing the data in Table 1.2 for these three sediment size groups. The data in Tables 1.1 and 1.2 are presented schematically in Figure 1.5 a, b.

The size distribution of the eroded (i.e. undispersed) sediment in Sample 1.18, which was taken from one of the small rills shown in Figure 1.3, is considerably coarser than the size distribution of the dispersed sediment, indicating that some of the sediment in the rill eroded as aggregates (Figure 1.5A). The eroded sediment contained some medium aggregates, which were comprised of sand, silt and clay material, and a small percentage of large aggregates which were comprised almost entirely of aggregated silt. The dispersed sediment was predominantly silt and clay with only 4% to sand. Two things are of note in these data. First, most of the clay in the dispersed sediment erodes as primary clay, with only small percentages incorporated in the aggregates. This is significant from the standpoint of the transport of agricultural chemicals and the subsequent pollution of receiving waters. Primary particles finer than 2 µm, which have the greatest pollution-carrying potential, are generally transported furthest along the sediment conveyance route, often reaching the permanent watercourse within a catchment. The second point to note is that the size distribution of both the eroded and dispersed sediment is significantly finer than that of the matrix soil, with considerable enrichment evident in both the silt and clay fractions. For the dispersed sediment, enrichment ratios were 1.5 and 2.0 for the clay and silt fractions respectively, with sand depleted by a ratio of 0.1. For the eroded sediment, clay was enriched by a ratio of 1.1 and silt by a ratio of 1.4 with sand depleted by 0.4. Even though some of the clay and silt was bound to aggregates, there was still more clay and silt in the eroded sediment than in the matrix soil. These data suggest that the flow in the rill channel at the time of sampling was only competent enough to transport the finer silt and clay fraction of the sediment, at the expense of the coarser sand fraction. The discharge and velocity data appear to support this interpretation of the selective or preferential erosion of fines.

TABLE 1.2 Size distributions of primary particles in three sediment size groups for the eroded sediment in Table 1.1

		Eroded sediment		Dispersed size distribution for each size class (%)			
	Size class	Percentage in size class	>63 µm	16–63 µm	2–16 µm	<2 µm
Sample 1.18	All sediment	100	3.9	29.9	57.2	9.0
	>250 µm	2.3	–	27.0	55.6	17.4
	63–250 µm	12.8	30.3	36.6	28.7	4.4
	<63 µm	84.8	–	31.1	59.5	9.4
Sample 2.18	All sediment	100	28.8	21.5	42.4	7.3
	>250 µm	7.8	68.8	7.7	19.3	4.2
	63–250 µm	30.8	76.1	8.2	12.8	2.9
	<63 µm	61.5	–	30.2	59.6	10.3
Sample 2.13	All sediment	100	37.1	26.6	31.6	4.7
	>250 µm	11.4	46.0	23.1	26.0	4.9
	63–250 µm	39.1	81.5	9.5	7.9	1.2
	<63 µm	49.5	–	42.3	50.3	7.4
Sample 3.13	All sediment	100	51.2	18.2	26.4	4.2
	>250 µm	15.6	64.5	15.5	17.2	2.8
	63–250 µm	49.2	83.4	8.2	7.3	1.2
	<63 µm	35.2	–	37.3	54.1	8.6
Sample 4.13	All sediment	100	40.4	21.1	33.5	4.9
	>250 µm	15.1	37.1	20.3	35.8	6.9
	62–250 µm	46.7	74.5	9.1	13.9	2.5
	<65 µm	38.2	–	35.5	56.3	8.3
Sample 5.13	All sediment	100	21.5	38.0	35.1	5.4
	>250 µm	22.1	36.8	29.0	30.0	4.2
	63–250 µm	31.4	42.6	24.2	28.1	5.1
	<63 µm	46.5	–	48.4	44.7	6.9

Figure 1.5a also shows the size distributions of eroded and dispersed sediment from one of the larger rills in sub-catchment A (Sample 2.18). These data were taken at site 2, also during the storm of 18 December 1992. As with Sample 1.18, the eroded sediment is coarser than the dispersed sediment, indicating that some of the sediment is ending as aggregates. However, the overall size distribution of the eroded sediment is significantly coarser than that in the small rill discussed previously, with considerably more medium and large aggregates being transported in the runoff (Table 1.2). This appeared to be related to the flow conditions in this larger rill which was fully competent to transport this coarser material. Mean discharge in the rill at the time of sampling was 6.2 l s^{-1} with a mean velocity 0.74 m s^{-1}, almost double that of the flow in the small rill (i.e. Sample 1.18,

(a)

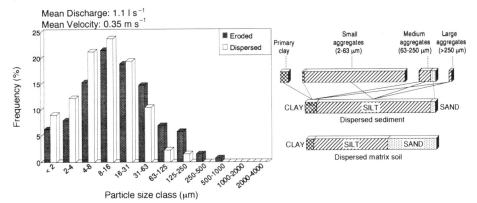

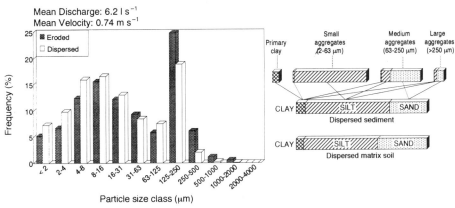

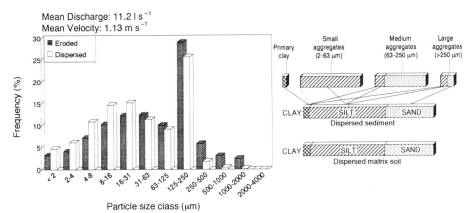

FIGURE 1.5 Size distribution of the eroded sediment and dispersed sediment for the six sediment samples in Table 1.1. The size distribution of primary particles in the three sediment size groups, along with the size distribution of the dispersed sediment and dispersed surface soil, are shown schematically on the right of each plot

(b) **SAMPLE 3.13: THALWEG RILL SUSPENDED SEDIMENT 13 January 1993**

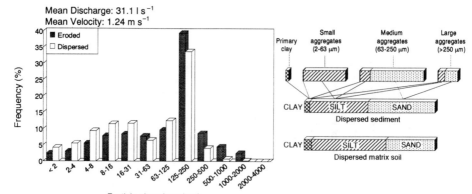

SAMPLE 4.13: FEEDER RILL SUSPENDED SEDIMENT 13 January 1993

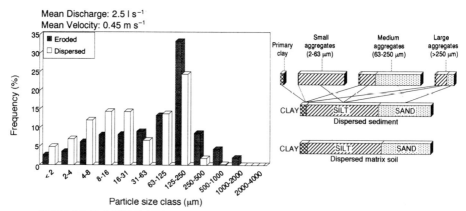

SAMPLE 5.13: DOWNSLOPE RILL SUSPENDED SEDIMENT 13 January 1993

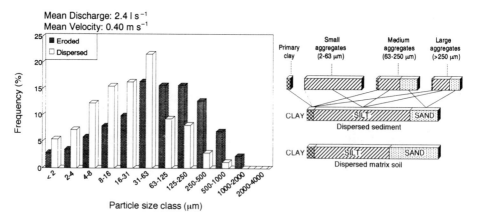

FIGURE 1.5 (*continued*)

Figure 1.5a). The size distribution of the eroded sediment and the dispersed sediment varied little from the size distribution of the dispersed surface soil. These sediment data therefore show almost no evidence of selectivity in particle sizes during runoff.

Data from the same rill (Site 2), but sampled during the storm on 13 January 1993 (Sample 2.13, Figure 1.5a), show a similar pattern of sediment eroding as aggregates but with a size distribution that is coarser than during the flow on 18 December 1992. About 40% of the sediment is eroding as medium-sized aggregates (versus 30% on 18 December) with about 12% eroding as large aggregates (versus 8% on 18 December). Flow conditions during the January storm were considerably more erosive than during the December storm, with mean rill discharges of 11.2 l s^{-1} and mean velocities of 1.1 to 1.2 m s^{-1} during sampling.

An important observation in the data for Sample 2.13 is the depletion of fines in the eroded sediment relative to the dispersed matrix soil, with depletion ratios of 0.5 for clay and 0.8 for silt. Sand-sized sediment is enriched in the eroded sediment by a factor of 1.3. Two mechanisms may explain the depletion of the clay and silt fractions in these samples. The first relates to flow conditions. It appears that the flow in the larger rill on 13 January 1993 is more competent to transport the coarser material than the flow on 18 December 1992, and in fact does so at the expense of the finer size fractions. A second possibility, however, is that there has been a depletion or exhaustion of fine material *between* the two storm events, and thus in terms of fine material, the flow is supply-limited.

The data from the main thalweg rill system, sampled at Sites 3, 4 and 5 on 13 January 1993, reveal similar and consistent patterns of eroded material being transported as aggregates. The data from the main rill channel (Sample 3.13, Figure 1.5b) show that almost 50% of the sediment was transported as medium-sized aggregates with 16% transported as large aggregates. Like Sample 2.13 from the same storm, the eroded sediment also showed the depletion of clay and silt relative to the dispersed matrix soil, with depletion ratios of 0.4 for day and 0.6 for silt. The fact that more aggregated material is being transported in the main rill channel, and that clay and silt have undergone depletion, can again be related to flow conditions. Mean discharge in the main rill was calculated at 31.1 l s^{-1} and mean velocities at 1.24 m s^{-1} although velocities did fluctuate markedly along the main rill, with rates of greater than 2 m s^{-1} measured at several locations. Thus, as with the rill system at Site 2, flow in the thalweg rill was fully competent at all times to transport the coarser material at the expense of the silt and clay fractions.

The sediment sizes in the thalweg rill, as outlined above, are also remarkably similar to those in the feeder rills (Sample 4.13, Figure 1.5b), taken at Site 4. Depletion ratios for silt and clay in the eroded sediment are almost identical to those in the main rill, although the medium and large aggregates contain slightly more silt than those in the main rill. Sample 5.13, taken in the thalweg rill near the stream at Site 5, also shows aggregated material being transported in the flow although the overall size distribution of the eroded sediment is markedly different from that in the feeder rills (4.13) and thalweg rill (3.13) upslope. Here, large aggregates make up 22% of the eroded sediment compared with about 15% for the thalweg rill and feeder rills upslope (see Table 1.2). There are, however, considerably less medium-sized aggregates in this downslope sample; 31% compared to about 47% for the upslope data. Despite the fact that there is significantly more silt incorporated in the large and medium aggregates, as shown by the dispersed sediment which is enriched in silt by a ratio of 1.4 relative to the dispersed matrix soil, the undispersed eroded sediment shows a depletion in clay and silt, with depletion ratios of 0.5 and 0.8 respectively. This further highlights the inaccuracies of using dispersed instead of undispersed particle-size data for agricultural soils. The dispersed sediment indicates that silt is enriched (by a ratio of 1.4) and that sand is depleted (by a ratio 0.6) relative to the matrix soil. Based on these dispersed data alone, it would be correct to infer that the flow here is selectively removing fine particles through the process of detachment and transport at the expense of coarse particles. The size distribution of the undispersed eroded sediment, however, suggests exactly the opposite. Both clay and silt are depleted in the eroded sediment relative to the matrix soil, as indicated above, with the sand-sized fraction, which is made of medium and large aggregates, enriched by a factor of 1.4.

The data presented in Figure 1.5 are summarized in Figure 1.6. The size distribution of the dispersed matrix soil and the size distributions of the six

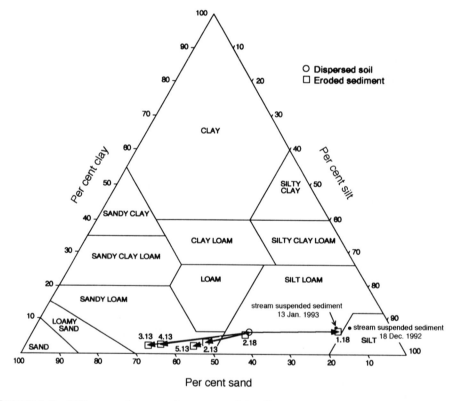

FIGURE 1.6 Differences between the texture of the dispersed surface soil and the "texture" of the resulting sediment in the undispersed form as it erodes in rill flow during the two rainstorms. The particle size distributions of the undispersed suspended sediment in the stream at peak discharge for the two storms are also plotted on the diagram

undispersed eroded samples have been plotted on a USDA Textural Classification Chart. We have also plotted the particle size distribution of the undispersed suspended sediment in the stream at peak discharge for the two storms. For Sample 1.18, rill flow was unable to transport the coarser fraction of the sediment available in the matrix soil and hence there is selective removal of fine particles at the expense of the coarse ones. The eroded sediment size distribution of Sample 2.18 is similar to that of the dispersed soil, indicating that the sediment eroding from the slopes is representative of the surface soil from which it was eroded and that little selectivity or enrichment resulted under these conditions. Samples from the 13 January storm (2.13–5.13) produced sediment that resembled sandy loams texturally. In all cases, flow was fully competent to transport the coarse material. In essence, flow in these rills "selectively" removes the coarser particles as aggregates at the expense of the finer fractions. The stream suspended sediment was considerably finer than the material eroding from the catchment slopes during the two storms, indicating that much of the slope-derived sediment is deposited along the conveyance route, in the present case as alluvial fans along field boundaries (see Figure 1.2).

USING SEDIMENT SIZE DATA

By using assumed size-transport capabilities for several flow conditions with the sediment size data from Sample 5.13, we can illustrate how flow conditions might affect the transportation of sediment (Figure 1.7). A very slow flow, assumed capable of carrying only particles smaller than 4 µm, could carry only about 6% of

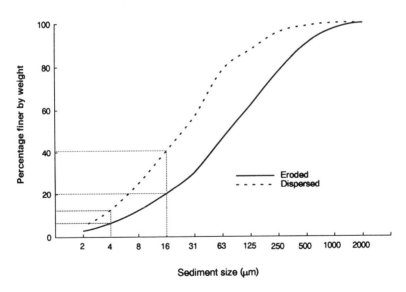

FIGURE 1.7 Cumulative frequency curve of the eroded and dispersed sediment from Sample 5.13, showing the difference in the percentages of sediment that could potentially be transported depending on whether eroded or dispersed data are used

the eroded sediment, although the size distribution of the dispersed sediment indicated that about 13% could be carried in the dispersed state. Similarly, for a flow that could transport all particles up to 16 µm, about 20% of the eroded sediment could be carried although the size distribution of the dispersed sediment indicated that 40% could be transported in the dispersed state. The percentages of sediment that could potentially be transported for different flow conditions thus changes greatly if the dispersed size distribution of the sediment is assumed to govern sediment transport instead of the effective, undispersed sediment size distribution, which includes aggregated sediment. Thus, large errors may result if the transportability of sediment is inferred from dispersed textural characteristics rather than the actual, *in situ* sediment sizes.

SUMMARY AND CONCLUSIONS

This study of the sediment size distributions resulting from runoff and erosion during natural rains gave the following results:

1. The size distribution of the undispersed eroded sediment was coarser than the size distribution of the sediment's primary particles, because much of the sediment was in the form of aggregates.
2. Most of the medium (63–250 µm) and large (>250 µm) aggregates contained sizeable amounts of silt-sized particles but only small amounts of clay, indicating that most of the clay erodes as primary clay. This may have important implications for sediment-bound pollutant transport because many of the soil's potential pollutants are associated with this component.
3. The size distribution of the undispersed eroded material was related to flow conditions. Flows with large discharges and velocities transported a larger percentage of coarser aggregates than less erosive flows.
4. Sediment leaving the catchment contained considerably more fine particles than that eroding from the catchment slopes, indicating that most of the eroded sediment was deposited along the conveyance route.
5. The dispersed sediment (and, for that matter, the dispersed soil) size distribution is not appropriate for determining the transportability of sediment eroded by overland flow. We have shown that large errors may result if the transportability of sediment is inferred from dispersed textural characteristics rather than the actual or effective sediment sizes.

ACKNOWLEDGEMENTS

We thank Maurice Velterop and John Boardman for their help with the field sampling and Chris Jackson for his help with the laboratory procedures.

REFERENCES

Abrahams, A. D. and Parsons, A. J. 1990. Determining the mean depth of overland flow in field studies of flow hydraulics. *Water Resources Research*, **26**, 501–503.

Abrahams, A. D., Parsons, A. J. and Luk, S.-H. 1986. Field measurement of the velocity of overland flow using dye tracing. *Earth Surface Processes and Landforms*, **11**, 653–657.

Alberts, E. E., Wendt, R. C. and Piest, R. F. 1983. Physical and chemical properties of eroded soil aggregates. *Transactions of the ASAE*, **26**, 465–471.

Foster, G. R., Young, R. A. and Niebling, W. H. 1985. Sediment composition for nonpoint source pollution analyses. *Transactions of the ASAE*, **28**, 133–139, 146.

Horton, A., Poole, E. G., Williams, B. J., Illing, V. C. and Hobson, G. D. 1987. *Geology of the Country Around Chipping Norton*. British Geological Survey, London.

Jarvis, M. G., Allen, R. H., Fordham, S. J., Hazelden, J., Moffat, A. J. and Sturdy, R. G. 1984. *Soil Survey of England Wales*, Bulletin 15, Harpendon Press.

Meyer, L. D., Harmon, W. C. and McDowell, L. L. 1980. Sediment sizes eroded from crop row sideslope. *Transactions of the ASAE*, **23**, 891–898.

Meyer, L. D., Line, D. E. and Harmon, W. C. 1992. Size characteristics of sediment from agricultural soils. *Journal of Soil and Water Conservation*, **47**, 107–111.

Ongley, E. D., Bynoe, M. C. and Percival, J. B. 1981. Physical and geochemical characteristics of suspended solids, Wilton Creek, Ontario. *Canadian Journal of Earth Science*, **18**, 1365–1379.

Parsons, A. J., Abrahams, A. D. and Luk, S.-H. 1991. Size characteristics of sediment in interrill overland flow on a semiarid hillslope, southern Arizona. *Earth Surface Processes and Landforms*, **16**, 143–152.

Slattery, M. C. and Bryan, R. B. 1992. Hydraulic conditions for rill incision under simulated rainfall: a laboratory experiment. *Earth Surface Processes and Landforms*, **17**, 127–146.

Walling, D. E. 1990. Linking the field to the river: sediment delivery from agricultural land. In Boardman, J., Foster, and Dearing, J.A. (eds), *Soil Erosion on Agricultural Land*, Wiley, Chichester, pp. 129–152.

Young, R. A. 1980. Characteristics of eroded sediment, *Transactions of the ASAE*, **23**, 1139–1142, 1146.

2

Dune Geometry and Sediment Transport: Fraser River, British Columbia

R. A. KOSTASCHUK AND S. A. ILERSICH

Department of Geography, University of Guelph, Ontario, Canada

ABSTRACT

Mean height, length and migration rates of a group of dunes in the Fraser River estuary, British Columbia, lagged behind changes in channel velocity associated with variations in tidal range, displaying a hysteresis effect. Detailed measurements over individual dunes revealed an absence of lee-side flow separation and lee-side slope angles much less than the friction angle of bed material. These characteristics were attributed to a predominance of bed-material movement in suspension rather than as bed load. A comparison of sediment transport estimates based on migrating dunes with direct measurements of bed-material load showed that approximately 5% of sediment trapped within migrating dunes was bed load and the remainder was transported in suspension. The mean contribution of measured bed load to total measured bed-material load was about 1%.

INTRODUCTION

Sediment in rivers and estuaries is transported as wash load and bed-material load. Wash load consists of fine material transported in continuous suspension and does not appear in appreciable quantities on the bed. Bed-material load is coarser bottom sediment transported in traction or saltation as bed load and in intermittent suspension (Murphy and Aguirre, 1985). Bed load moves within a few grain diameters of the bed and suspended bed-material load is diffused vertically in the water column. Almost as soon as sandy bed material begins to move it is moulded into flow-transverse bedforms of various shapes and sizes. It is generally agreed (e.g. Ashley, 1990) that small-scale bedforms (<0.1 m in height, <0.5 m in length) that are not correlated with flow depth can be termed "ripples". A confusing variety of terms, however, has been used to describe large-scale bedforms (e.g. dunes, sand waves, megaripples) and Ashley (1990) recommends that a single term, "dune", be used; this nomenclature is adopted here.

River Geomorphology. Edited by Edward J. Hickin
© 1995 John Wiley & Sons Ltd

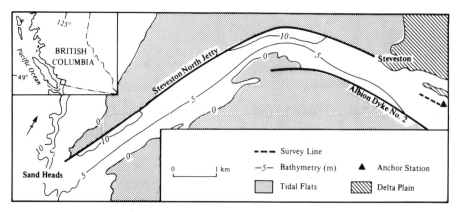

FIGURE 2.1 Lower Fraser River estuary showing survey line and anchor station near Steveston

Many studies in a variety of environments have examined the geometry and migration rates of dunes and used dune movement to estimate bed-material transport rates (e.g. Bokuniewicz et al., 1977; Langhorne, 1982; Aliotta and Perillo, 1987; van den Berg, 1987; Kostaschuk et al., 1989a). Most of these investigations make the assumption that such estimates reflect bed-load transport, although this has never been tested in the field. The purpose of this investigation is to examine relations between dune migration and sediment transport in Fraser River estuary near Steveston, British Columbia (Figure 2.1). The study has two distinct objectives: (1) to examine temporal changes in mean group-characteristics of dunes in the dune field, and (2) to compare estimates of sediment transport based on individual migrating dunes with direct measurements of bed-material load over the same dunes.

SETTING

Fraser River drains 250 000 km^2 of mountainous terrain on the west coast of Canada and has a mean annual discharge at the mouth of 3400 m^3 s^{-1}. Discharge is low in the winter, rising rapidly in spring in response to snowmelt, and declining gradually in the summer to low flow in the autumn. Tides in the estuary are semi-diurnal with distinct inequalities in the heights of the two high and low tides each day. The mean tidal range is approximately 3 m, with spring tides approaching 5 m and neap tides 2 m. During low river discharge and spring tides, mixing in the estuary is enhanced and the channel is only moderately stratified (Hodgins et al., 1977). High river discharges restrict mixing and the estuary becomes a highly stratified salt-wedge system. The salt-wedge migrates along the channel, its position determined by river discharge and tidal height (Kostaschuk and Atwood, 1990).

A bottom sampling programme by Kostaschuk et al. (1989a) showed that the mean grain size (Folk and Ward, 1957) of bed material in the estuary was around 0.25 mm and that it varied little seasonally. Measurements of suspended-sediment transport over tidal cycles during a variety of river discharges (Kostaschuk and Luternauer, 1989; Kostaschuk et al., 1989b) indicate that sediment movement in the estuary is controlled by river discharge, tidal height and flow stratification. Bed-material transport is overwhelmingly downstream-directed and is greatest under conditions of high river discharge, ebb tides and unstratified flow. Upstream-directed currents occur during flood tides, but near-bed velocities are low and there is no evidence for significant upstream bed-material transport.

Dunes in Fraser Estuary vary in length from 4 m to greater than 100 m and in height from 0.3 m to greater than 5 m (Kostaschuk *et al.*, 1989a), making them medium to very large dunes in the Ashley (1990) scheme. Multitrack surveys of some bedforms larger than 10 m in length (Kostaschuk and MacDonald, 1988) reveal a curved, concave-downstream planform with crests that are continuous for at least 300 m across the channel. This planform geometry indicates a two-dimensional shape (Ashley, 1990), at least for the larger features. Kostaschuk *et al.* (1989a) monitored bi-weekly variations in dune characteristics and found that changes in dune height and length follow, but lag behind, seasonal variations in river discharge.

METHODS

Field data were collected in Fraser Estuary near Steveston (Figure 2.1) between 30 May 1989 and 6 July 1989, using the survey launch *CSL Jaeger* in a period which included the largest freshest discharges of the year (Table 2.1). Preliminary surveys showed that large dunes (>10 m in length) had developed and these were measured during a 2 to 3 hour interval about the lowest low tide of the day (Table 2.1). This interval was chosen because flow is unstratified, estuary currents are strongest and relatively steady (limited tidally-induced flow acceleration or deceleration), most bed material is transported during this period, and dunes are most active (Kostaschuk and Luternauer, 1989; Kostaschuk *et al.*, 1989a, 1989b).

Bedform geometry and migration rates were measured with a 200 kHz Apelco echosounder. Temporal changes in the mean characteristics of the dune field in the Steveston reach were evaluated from soundings run along a 530 m survey line near the centre of the channel (Figure 2.1). Navigation beacons were used for along-stream positioning and range markers on the north side of the channel for cross-stream positioning. For detailed profiles of individual dunes and associated sediment transport predictions, very precise soundings were obtained by anchoring the vessel to a navigation buoy (Figure 2.1) with a 200 m cord marked at 0.5 m intervals. The cord was slowly played out as the sounder was running and position fixes made as the vessel passed over the dune. Soundings for both the dune field and individual profiles were obtained at the beginning and end of the 2 to 3 hour time interval to estimate dune migration rates.

Horizontal-current speed and direction, suspended-sediment concentration and near-bed transport were measured at five or six positions along each bedform, located using the marked cord, to directly measure sediment transport over the individual dunes surveyed. A Marsh McBirney 527 electromagnetic flow meter was used to measure current velocity, a pump sampler for suspended-sediment samples, and a Helley-Smith sampler for near-bed transport. Velocity and suspended-sediment concentration were measured at a number of positions above the bed, starting at 0.47 m, to provide velocity and concentration profiles at each location. Velocity was time-averaged over 2 minutes at each position and concentration samples usually took about 30 s each to collect. The suspended sediment samples were filtered through 0.45 μm filters to determine sediment concentration. A 4 litre pump sample was taken at 0.47 m above the bed for grain-size analysis of the suspended load, using the bottom withdrawal method (e.g. McCave, 1979). A single 2–5 minute Helley-Smith sample was taken after velocity and concentration profiles were completed. A sample of the bed at each individual dune surveyed was obtained with a dredge bottom sampler. Grain-size distributions of bottom samples were determined with an automated sedimentation column (e.g. Rigler *et al.*, 1981).

Suspended bed-material concentration was determined by subtracting the washload concentration from the total suspended concentration, using grain-size distributions from the 4 litre samples taken 0.47 m from the bed (Figure 2.2). McLean and Church (1986) suggest that suspended sediment less than 0.125 mm is wash load in the Fraser. Recent data indicate, however, that 0.200 mm may be more appropriate (M.A. Church, pers. comm., 1991). We used both of these values to determine wash concentration, then subtracted these from total concentration at each level above the bed to determine bed-material concentration. This approach is based on the reasonable assumption that wash load is evenly distributed vertically throughout the flow.

Echosounding profiles of dunes were affected by positioning errors and limitations of the sounder. A field test using standard surveying techniques showed that positions obtained with shore markers are usually replicable to within 2 m.

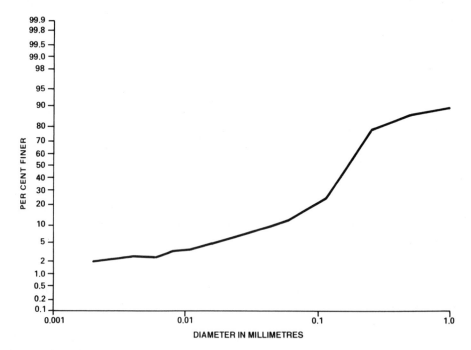

FIGURE 2.2 Grain-size distribution of suspended sediment 0.47 m from the bed on 21 June 1989, determined by the bottom withdrawal method

Horizontal positions using the marked cord have an error of ±0.5 m, the precision of the marked cord. For the depths encountered in this study, the sounder is accurate to within ±0.1 m vertical depth (manufacturer's specification).

Calculation of suspended bed-material load is based primarily on measurements of velocity and bed-material concentration. The current meter is accurate to ±2% (manufacturer's specification) so current meter error was minimal. Sediment concentration is subject to error associated with the pump sampler, filtration procedures and grain-size estimates from the bottom withdrawal method. The bias of the pump sampler was assessed by applying a linear regression model to 22 simultaneous concentrations from the pump and a USP-61 point-integrating river sampler. The USP-61 sampler is regarded as the standard for point-integrated river sampling and is assumed to have a smaller error than the pump. The regression is highly significant ($p = 0.0001$), has a high level of explanation ($R^2 = 0.969$), a low root mean square error (32.9 mg l^{-1} and a slope close to 1 (1.019). These results indicate that the pump sampler provides an excellent approximation of USP-61 samples. The variability associated with final calculations of suspended bed-material load, however, remains unknown.

Point Helley-Smith samples can vary considerably because of temporal variations in sediment transport and trapping efficiency of the sampler (Ludwick, 1989; Gomez et al., 1990). In order to assess this variability, we took 19 point samples at a single position near a dune crest over a 60 minute period on 7 July. Each sample was 2 minutes in duration. The samples had a mean transport rate of 0.031 kg m^{-1} s^{-1} and a standard deviation of 0.016 kg m^{-1} s^{-1}. The 99% confidence limits were 0.020–0.042 kg m^{-1} s^{-1}, or ±35% about the mean.

DUNE-FIELD CHARACTERISTICS

Echosoundings along the 530 m survey line (Figure 2.1) revealed temporal changes over the survey period in the central dune field near Steveston (Figure 2.3). Dune height, length, steepness (height/length) and migration rate were measured for each dune and these data were averaged for each of the 19 sets of soundings (Table 2.1). Individual dunes were easily recognized on soundings before and after the 2 to 3 hour sampling interval (Figure 2.3), allowing determination of migration rates. Measurements of dune length and migration rate are affected by errors in navigation positioning and measurements of dune height are affected by the precision of the sounder. Comparisons of dune migration rates must be treated with some caution because the positioning error of 2 m is close to the slowest dune movements over the sampling interval.

Correlation analysis (0.05 significance level) was used to assess relations between dune-field characteristics (Table 2.1). Ashley (1990) has shown that dune height and length are related as a power function, so we applied a power correlation model to this relation. Since we have no reason to expect non-linearity in the remaining relations, only linear models were applied (Table 2.2a). All of the correlations involving dune steepness are weak and statistically insignificant, probably because of the limited variation in the steepness data. The remaining relations are moderately strong and statistically significant. Height and length are positively correlated as

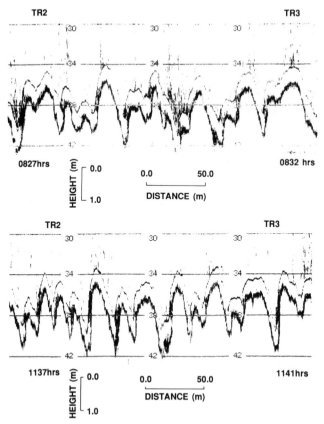

FIGURE 2.3 Echosounding profiles along the 530 m survey line (Figure 2.1) for 15 June 1989. The true bed surface is the thin trace at the top of the profiles. The top profile was obtained before the survey and the lower profile after. TR2 and TR3 are position fixes from shore markers; chart scale is in feet. Downstream is to the left. The "dirty" nature of the records is due to high suspended-sediment concentrations

expected. Both height and length correlate negatively with migration rate, indicating that as dune size increases, migration rate decreases.

The influence of flow conditions on dune characteristics was assessed by examining correlations among low-tide mean cross-sectional flow velocity and dune height, length, steepness and migration rate (Table 2.2b). Mean cross-sectional velocity is a prediction from the Fraser River Mathematical Model (Ages and Woolard, 1976) which integrates river discharge and tidal height. The model does not adequately reflect velocity structure during stratified flows in the Fraser but works extremely well for unstratified conditions such as those encountered in this study (Ages and Woolard, 1976). It was not possible to use measured flow-velocity in this analysis because measured data were not available on all occasions the dune field was monitored (Table 2.3). Predicted velocities were averaged over the 2 to 3 hour sampling period to provide a value that reflected conditions during the period of dune activity. During our study period there was a limited range in river discharge,

TABLE 2.1 Dune-field characteristics and estuary flow conditions

Date	$H(m)$	$L(m)$	H/L	$U_{bm}(m^{-1})$	$T(m)$	$Q(m^3 s^{-1})$	$U_p(m s^{-1})$
1 June	1.10	22.6	0.05	0.0016	0.6	4640	1.40
6 June	1.25	35.8	0.03	0.0005	0.4	6580	1.50
7 June	1.20	36.5	0.03	0.0011	0.7	6640	1.33
8 June	1.37	35.0	0.04	0.0006	1.1	6840	1.18
14 June	0.86	23.6	0.04	0.0019	1.4	6810	1.15
16 June	1.28	22.3	0.06	0.0012	0.9	6780	1.30
19 June	1.85	37.0	0.05	0.0013	1.1	6790	1.45
20 June	2.10	35.7	0.06	0.0010	0.4	6480	1.35
21 June	2.20	37.2	0.06	0.0008	0.4	5960	1.38
23 June	1.22	39.5	0.03	0.0016	0.8	5290	1.25
27 June	1.56	35.6	0.04	0.0012	1.7	4910	1.20
28 June	1.84	42.2	0.04	0.0002	1.2	4890	1.30
29 June	1.46	36.1	0.04	0.0004	0.8	4980	1.45
30 June	1.19	28.5	0.04	0.0011	0.5	5130	1.40
1 July	0.38	34.7	0.05	0.0009	0.3	4980	1.55
2 July	1.54	34.0	0.04	0.0004	0.2	4720	1.60
3 July	1.60	38.9	0.04	0.0003	0.2	4510	1.60
4 July	1.70	39.7	0.04	0.0003	0.4	4430	1.55
5 July	1.91	42.5	0.04	0.0002	0.6	4430	1.58

Notes: Mean dune height (H), length (L), steepness (H/L) and migration rate (U_{bm}) along the 530 m survey line. Fraser River discharge at Hope (Q) (135 km upstream of the river mouth at Sand Heads), lower low water tidal height at Point Atkinson (T) (30 km north of the river mouth at Sand Heads) and mean cross-sectional channel velocity at Steveston (U_p) for the 1989 study period

TABLE 2.2 Correlation analyses

Correlation	r	p
(a) H vs. L	0.66	0.002
H vs. H/L	0.46	0.057
H vs. U_{bm}	−0.52	0.023
L vs. H/L	−0.34	0.156
L vs. U_{bm}	−0.66	0.002
H/L vs. U_{bm}	−0.15	0.531
(b) H vs. U_p	0.37	0.127
L vs. U_p	0.29	0.231
H/L vs. U_p	0.16	0.828
U_{bm} vs. U_p	0.58	0.016
(c) G_{sc} vs. U_p	0.24	0.422

Notes: Pearson product-moment correlation coefficients (r) and probability of chance occurrence (p) for (a) mean dune characteristics of height (H), length (L), steepness (H/L) and migration rate (U_{bm}) along the 530 m survey line ($n=19$); (b) mean dune characteristics along the 530 m survey line and mean cross-sectional velocity (U_p) ($n=19$); (c) mean cross-sectional velocity and sediment transport within dunes (G_{sc}) ($n=14$)

and cross-sectional velocity was controlled primarily by variations in tidal height (Table 2.1). The inverse relation between tidal height and mean cross-sectional velocity is due to the steeper water-surface slopes at lower tidal elevations and the greater overall tidal fall and flow acceleration during the lowest tides.

The only statistically significant result in the correlations of Table 2.2b is for dune migration rate, although the relation is rather weak. Kostaschuk *et al.* (1989a) found weak relations between dune characteristics and river discharge over a seasonal time frame and they attributed these to hysteresis effects. We examined the potential for hysteresis by plotting mean cross-sectional velocity versus dune height, length and migration rate on phase diagrams (Figure 2.4). The phase diagrams showed counter-clockwise hysteresis loops in all relations in June, with changes in bedform

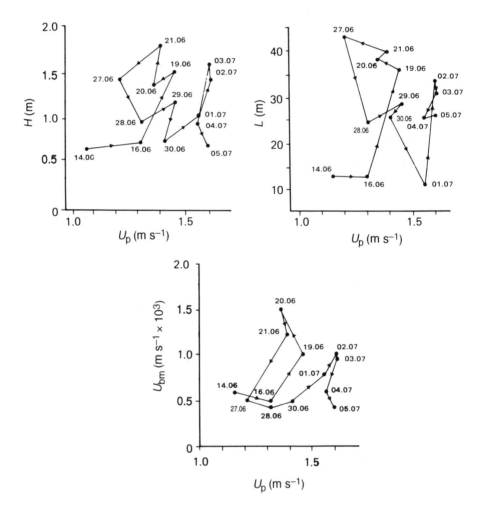

FIGURE 2.4 Phase diagrams illustrating hysteretic relations between mean cross-sectional velocity (U_p) and dune height (H), length (L) and migration rate (U_{bm}). The numbers at data points indicate dates (day.month) and the arrows indicate time direction

characteristics lagging behind changes in velocity. Hysteresis is poorly developed in July. The discharge hysteresis loops described by Kostaschuk *et al.* (1989a) are also counter-clockwise. Hysteresis clearly occurs over relatively short periods when variations in velocity are controlled by tides, as well as seasonally when the strongest controls are exerted by changes in river discharge (Kostaschuk *et al.*, 1989a).

Allen (1983) provides an explanation for bedform hysteresis that applies to Fraser River. Bedform populations are affected by creation–destruction processes, driven by sediment transport, in which new forms replace older ones. A group of bedforms thus responds to changing flow velocity by changing its size composition, with the rate of response inversely related to bedform size and directly related to current speed (Allen, 1983). During this investigation in Fraser Estuary, changes in velocity were due primarily to tidal variations. The period of low tides between 14 and 24 June 1989 resulted in an increase in velocity to which the dunes did not adjust until tidal height began to increase, and velocity to decrease, later in the month. Dunes were larger during this period and velocities relatively low, resulting in broad hysteresis loops between 16 and 30 June. The weaker hysteresis and faster response between 30 June and 5 July occurred because the dunes were smaller and flow velocities were higher.

SEDIMENT TRANSPORT ESTIMATES BASED ON DUNE MIGRATION

Simons *et al.* (1965) provide a method for computing sediment transported within a migrating bedform, G_{sc}:

$$G_{sc} = \sigma(1 - P)\beta H U_{bm} \qquad (2.1)$$

in which G_{sc} is the dry-weight transport rate per unit width of channel, σ is sediment density, β is a bedform shape factor, P is porosity, H is bedform height and U_{bm} is bedform migration rate.

Sediment transport estimates based on dune migration were obtained using equation (2.1) and data from surveys of 14 individual dunes obtained as the vessel

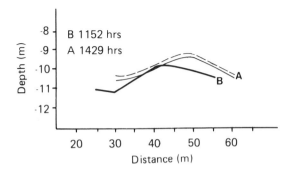

FIGURE 2.5 Profiles of a dune before (B) and after (A) a survey on 21 June 1989. Positions were maintained using a marked cord anchored to a navigation buoy. The Fraser River Mathematical Model (Ages and Woolard, 1976) was used for tidal adjustments to the "after" profile (broken line)

TABLE 2.3 Individual dune characteristics and sediment transport

Date	H (m)	L (m)	H/L	U_{bm} (m s^{-1})	D_{50} (mm)	G_{sc} (kg m^{-1} s^{-1})	H-S (kg m^{-1} s^{-1})	sand >0.125 mm (kg m^{-1} s^{-1})	sand >0.200 mm (kg m^{-1} s^{-1})
14 June	0.65	12.5	0.05	0.0006	0.29	0.37	0.034		
16 June	0.70	12.5	0.06	0.0005	0.30	0.33	0.036		
19 June	1.50	36.5	0.04	0.0010	0.33	1.39	0.070	9.97	5.41
20 June	1.35	38.0	0.04	0.0015	0.29	1.98	0.113	12.88	10.60
21 June	1.80	40.0	0.04	0.0012	0.31	2.10	0.168	13.24	6.13
27 June	0.40	43.5	0.03	0.0005	0.28	0.71	0.017	10.89	4.63
28 June	0.95	24.1	0.04	0.0005	0.32	0.48	0.039	2.28	1.00
29 June	1.16	28.0	0.04	0.0006	0.31	0.68	0.027	5.04	1.71
30 June	0.74	26.8	0.03	0.0005	0.32	0.38	0.054	6.85	2.02
1 July	1.01	11.5	0.09	0.0008	0.30	0.76	0.103		
2 July	1.50	33.1	0.05	0.0010	0.30	1.43	0.122	5.42	1.30
3 July	1.62	31.8	0.05	0.0009	0.31	1.34	0.037	5.12	1.69
4 July	0.90	25.5	0.04	0.0006	0.29	0.53	0.094		
5 July	0.69	26.8	0.03	0.0004	0.28	0.27	0.053		

Notes: Individual dune height (H), length (L), steepness (H/L) and migration rate (U_{bm}). Median bed-material grain size (D_{50}), sediment transport within dunes (G_{sc}), mean Helley-Smith near-bed transport rates averaged over the dune (H-S), and mean suspended bed-material load larger than 0.125 mm (sand >0.125 mm) and larger than 0.200 mm (sand >0.200 mm) are averaged over the dune

was anchored to the navigation buoy (Figure 2.1). These dunes were bankward extensions of dunes in the dune field (Figure 2.3). Figure 2.5 is an example of individual dune profiles before and after a 3 hour survey. Because tidal height usually increased over this time interval, depth adjustment was required. Some dunes changed form slightly during migration, but most maintained their profile shape.

Table 2.3 summarizes dune characteristics, median grain size and sediment transport estimates for the individual bedforms. Dune heights and lengths were slightly less than mean values from the dune field in the centre of the channel, because the channel was shallower at the navigation buoy. Dune steepness and migration rates were comparable to those from the central dune field, with the exception of those of 1 July when the measured dune had a steepness of 0.09. In general, data from individual dunes (Table 2.3) show good agreement with data from the central dune field (Table 2.1) for the same days.

In the transport estimates of Table 2.3, sediment density was assumed constant at $\sigma = 2650$ kg m^{-3}. van den Berg (1987) reviewed the literature and found porosity $P = 0.4$ and the shape factor $\beta = 0.6$ in most studies; these values were adopted here. Results from the central dune field revealed hysteresis effects between flow velocity and dune characteristics (Figure 2.4). Since the transport equation is based on

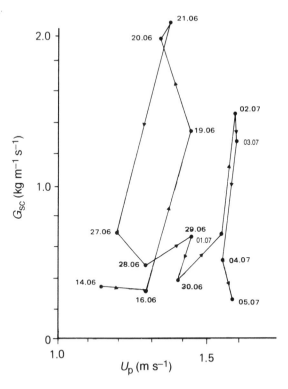

FIGURE 2.6 Phase diagram illustrating hysteresis between estimates of sediment transport from equation 2.1 (G_{sc}) versus mean cross-sectional velocity (U_p). The numbers at data points indicate dates (day.month) and the arrows indicate time direction. Transport rates are based on surveys of individual dunes (e.g. Figure 2.5)

bedform parameters, similar effects might exist between transport estimates and flow. We tested this hypothesis by correlating G_{sc} with mean cross-sectional velocity (Table 2.2c) and by plotting a phase diagram (Figure 2.6). The correlation was very low and statistically insignificant. The phase diagram displays patterns nearly identical to those for dune characteristics (Figure 2.4). This indicates that hysteresis in dune characteristics causes similar relations between flow and sediment transport estimates based on dune properties. This does not imply hysteresis between flow and bed-material transport at this time-scale, although it clearly complicates empirical modelling of sediment transport using transport rates based on dune surveys (e.g. Kostaschuk et al., 1989a).

MEASURED BED-MATERIAL TRANSPORT OVER DUNES

Flow velocity, suspended-sediment concentration and near-bed transport were measured over the 2 to 3 hour survey interval at five or six positions over individual dunes as the vessel was anchored to the navigation buoy. Measurements were first made in the trough at the upstream end of the dune and proceeded downstream to the next trough. Velocity and suspended-sediment measurements were obtained first at each position, followed by a Helley-Smith near-bed sample. Complete data sets were collected for 9 dunes, with partial sets for the rest (Table 2.3). Mechanical failure of the pump sampler was responsible for gaps in the suspended-sediment data.

Figure 2.7 is an example of velocity and total suspended-sediment concentration over a dune in the Fraser. Each vertical profile on a dune was divided into a number of segments, the lowest extending to 0.08 m above the bed, the maximum height reached by the Helley-Smith sampler. Bed-material load for each segment was calculated as the product of velocity, segment thickness and bed-material concentration and the segment values were summed to provide the bed-material loads for each location on the dune. Data at 0.47 m is taken to represent suspended bed-material load for the lowest segment. Bed-material concentration was determined for both the 0.125 mm and 0.200 mm wash-load boundaries and associated loads were expressed as sand >0.125 mm and sand >0.200 mm. These were then averaged over each dune (Table 2.3). Helley-Smith transport rates were also averaged over each dune to provide a mean near-bed transport rate (Table 2.3).

Traditional models of flow over dunes (e.g. Dyer, 1986) include flow separation on the lee side, yet none of the individual dunes that we examined show any evidence of flow separation (e.g. Figure 2.7). The current meter is equipped with a compass and 4 probes and should be able to detect upstream-directed currents in a separation zone. It is possible that the current meter was too far from the bed (0.47 m) to detect flow separation in some cases, but this explanation is unlikely to apply to the larger dunes. Dune morphology also suggests that flow separation may not develop. Dyer (1986) suggests that dunes with steepness (height/length) values greater than 0.06 should experience flow separation. With the exception of the small dune on 1 July, the individual dunes in the Fraser had steepness values less than 0.6. In addition, all lee-side slope angles were much less than the friction angle of the sediment (a maximum of 20° on 20 June: Figure 2.7), making flow separation unlikely (Dyer, 1986).

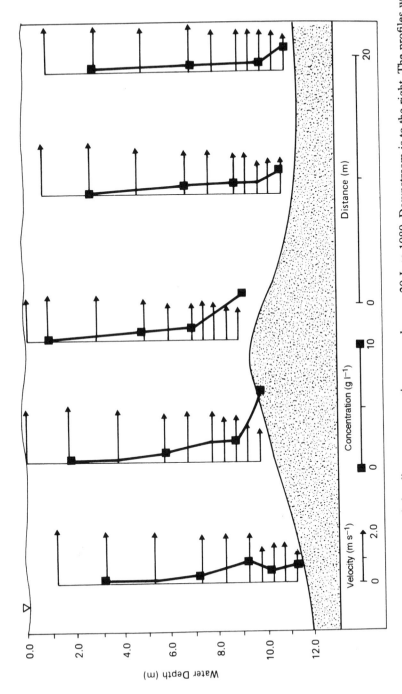

FIGURE 2.7 Velocity and total suspended-sediment concentration over a dune on 20 June 1989. Downstream is to the right. The profiles were obtained over a 3 hour period starting at the upstream end and positions were maintained using a graduated cord anchored to a navigation buoy

A possible explanation for the lack of flow separation and low lee-side slope angles of Fraser dunes is suggested by field investigations in Columbia River by Smith and McLean (1977) and McLean and Smith (1979). Columbia River is similar in size to the Fraser and bed-material characteristics are comparable. When flow strengths are low in Columbia River, bed-load transport dominates the bed-material load and dunes have long, gently sloping stoss sides and short, steep lee sides. Lee faces are usually inclined at the friction angle of the sediment (>28°) and characterized by a flow separation bubble. Sediment is transported as bed load up the stoss side and avalanches down the lee side. During higher flow strengths, bed-material transport is dominated by suspension. Suspended-sediment deposition "fills in" the lee-side separation zone, resulting in the loss of the flow separation bubble, stoss and lee sides that are of more comparable length, and lee-side slope angles that are much less than the friction angle of the sediment. Table 2.3 shows that most of the measured bed-material transport in Fraser Estuary was transported in suspension (sand >0.125 mm, sand >0.200 mm), indicating that the Columbia River hypothesis holds for Fraser dunes as well.

COMPARISON OF TRANSPORT ESTIMATES BASED ON DUNE MIGRATION WITH MEASURED BED-MATERIAL LOAD

Figure 2.8 summarizes relations between estimates of sediment transport based on dune migration (G_{sc}) Helley-Smith and suspended bed-material load. Considerable scatter exists in these relations, although the scatter is consistent over the range of data. On average, G_{sc} was larger than the Helley-Smith samples by a factor of 13, less than sand >0.125 mm by a factor of 8.7 and less than sand >0.200 mm by a factor of 4.2.

Although the differences between G_{sc} and Helley-Smith and suspended bed-material loads are large, some of these differences, plus the scatter in Figure 2.8, are due to sampling error. Such errors, however, will be reflected in the data. All of the Helley-Smith transport rates are lower than the values of G_{sc} (below the parity line of Figure 2.8) and all of the sand >0.125 mm and sand >0.200 mm values are larger than G_{sc} (above the parity lines of Figure 2.8). These results indicate that the Helley-Smith transport rates are significantly lower than G_{sc} and suspended bed-material loads are significantly higher than G_{sc}. For the rather small sample sizes represented here, the differences in transport rates have not been obscured by sampling error. The only data that experience overlap are sand >0.125 mm and sand >0.200 mm. We assessed the significance of the difference between these data (Table 2.3) with the Wilcoxon Signed Rank test at the 0.01 significance level. The only assumption of this test is that each pair of observations is independent of other pairs, a condition satisfied by the Fraser data. We expect sand >0.200 mm data to be smaller on average than sand >0.125 mm because of the larger wash-load boundary in the former. The probability associated with the test ($p = 0.004$) is less than the significance level, indicating that the average difference between sand >0.125 mm and sand >0.200 mm is significantly different from zero. Thus sampling error did not obscure the expected difference between the two measures of suspended bed-material load.

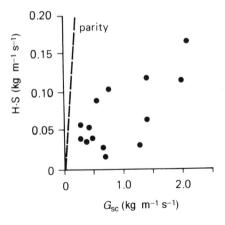
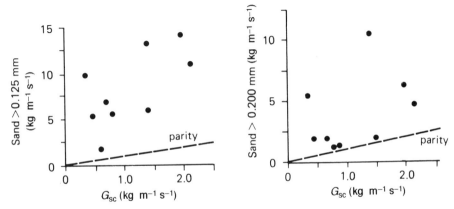

FIGURE 2.8 Scattergrams of sediment transport estimated from migrating dunes (G_{sc}) versus near-bed transport measured with a Helley-Smith sampler (H-S) and suspended bed-material loads (sand >0.125 mm, sand >0.200 mm)

The only direct field-estimates of sampling variability we have are the replicate measures of the Helley-Smith sampler. The 99% confidence limits of the Helley-Smith replicates were ±35% about the mean and the mean Helley-Smith transport rate from Table 2.3 is 0.069 kg m^{-1} s^{-1}. If we assume that the replicate confidence limits apply to the data on Table 2.3, then the upper confidence limit for the mean of Table 2.3 is 0.093 kg m^{-1} s^{-1}. This value is much less than any of the G_{sc} measurements, supporting the conclusion that sampling error cannot account for the large differences observed between Helley-Smith transport and G_{sc}.

In an environment where significant amounts of bed material are transported in suspension, a Helley-Smith sample represents bed load plus suspended bed material moving within the lowest 0.08 m of the flow (the height of the sampler aperture), since by definition bed load travels within a few grain diameters of the bed. For the particle-size range of bed sediment in the Fraser, field tests (Emmett, 1980) show that the Helley-Smith oversamples by approximately 150%. If we accept this oversampling rate as realistic for the Fraser, then "true-" bed load will be approximately 67% of the Helley-Smith rate. Accordingly, the data of Table 2.3 indicate that bed load comprised an average of about 5% of the sediment moving within the dunes. It follows that 95% of the sediment trapped within the migrating dunes was transported in suspension. Our measured data also provide an estimate of the percentage bed load of the total measured bed-material load during the high transport conditions that we examined in the Fraser. Again assuming that true bed load is around 67% of Helley-Smith samples, the data of Table 2.3 indicate that bed load is 0.6% of the total bed-material load when the 0.125 mm washload criterion is used and 1.2% when the 0.200 mm criterion is used.

Dune migration rates measured by Kostaschuk et al. (1989a) over 12 and 24 hour periods (one or two tidal cycles) range between 0.0001 and 0.0005 m s^{-1}, values comparable to the lowest estimates in this study (Table 2.1, Table 2.3). Most suspended sediment in Fraser Estuary is transported during the 2 to 3 hour period around low tide (Kostaschuk et al., 1989b) when dunes are actively migrating (Kostaschuk and Church, 1993). These observations suggest that most bed material is transported around low tide and bed load makes a very minor contribution to overall bed-material transport. Flows associated with lower river discharges and lower tidal ranges at other times of the year would likely have higher bed-load percentages, but these would be a small part of the annual bed-material load.

Our results have important implications for estimates of bed-load transport rates based on dune migration. The estimates of bed load for Fraser Estuary by Kostaschuk et al. (1989) were based on dune migration over 12–24 hour intervals. This time-frame includes periods of weak currents when the bed-load proportion of total bed-material load is relatively high, as well as periods of maximum suspended-sediment transport. The results of this study suggest, however, that the bed-load transport rates of Kostaschuk et al. (1989) are probably at least an order of magnitude too large. Dune-based estimates of bed load should be avoided in any environment where the dominant mechanism of bed-material transport is suspension.

CONCLUSIONS

The following conclusions can be drawn from this study of dunes in the Fraser Estuary:

1. Mean height, length and migration rate of dunes in a dune field lagged behind changes in channel velocity associated with variations in tidal range, displaying a hysteresis effect. Sediment transport estimates based on dune properties also lagged behind flow velocity.
2. Detailed measurements of individual dunes showed an absence of lee-side flow separation and lee-side slope angles much less than the friction angle of bed material. These results are attributed to a predominance of bed-material movement in suspension in the estuary.
3. A comparison of sediment transport estimates based on migrating dunes with direct measurements of bed-material load over the same dunes showed that approximately 5% of sediment trapped within migrating dunes was bed load and

the remainder was deposited from suspension. The mean contribution of measured bed load to total measured bed-material load was around 1%.

ACKNOWLEDGEMENTS

We thank Ray Sanderson for piloting the *Jaeger*. Special thanks are due to J.L. Luternauer for his continued support and M.A. Church for technical assistance. The manuscript greatly benefited from reviews by J. Luternauer, C. Amos, M.A. Church, K. Rood and M. Lapointe. Financial support was provided by a Natural Sciences and Engineering Research Council Operating Grant to R.A. Kostaschuk and J. Luternauer's Geological Survey of Canada Project 860022.

REFERENCES

Ages, A. and Woolard, A. 1976. Tides in the Fraser Estuary. Institute of Ocean Sciences, Sidney, BC, Pacific Marine Science Report 76–5.
Aliotta, S. and Perillo, G. M. E. 1987. A sand wave field in the entrance to Bahia Blanca Estuary, Argentina. *Marine Geology*, **76**, 1–14.
Allen, J. R. L. 1983. River bedforms: progress and problems. *International Association of Sedidmentologists, Special Publication* No. 6, pp. 19–33.
Ashley, G. M. 1990. Classification of large-scale subaqueous bedforms: a new look at an old problem. *Journal of Sedimentary Petrology*, **60**, 160–172.
Bokuniewicz, H. J., Gordon, R. B. and Kastens, K. A. 1977. Form and migration of large sand waves in a large estuary, Long Island Sound. *Marine Geology*, **24**, 185–199.
Dyer, K. R. 1986. *Coastal and Estuarine Sediment Dynamics*. Wiley-Interscience, Toronto.
Emmett, W. W. 1980. A field calibration of the sediment-trapping characteristics of the Helley Smith bed load sampler. *United States Geological Survey, Professional Paper* **1139**.
Folk, R. L. and Ward, W. C. 1957. Brazos River bar: a study in the significance of grain-size parameters. *Journal of Sedimentary Petrology*, **27**, 3–26.
Gomez, B., Hubbell, D. W. and Stevens, H. H. 1990. At-a-point bed load sampling in the presence of dunes. *Water Resources Research*, **26**, 2717–2731.
Hodgins, D. O., Osborn, T. R. and Quick, M. C. 1977. Numerical model of stratified flow. *American Society of Civil Engineers, Journals of the Waterways, Ports, Coastal and Ocean Division*, WW1, pp. 25–42.
Kostaschuk, R. A. and Atwood, L. A. 1990. River discharge and tidal controls on salt-wedge position and implications for channel shoaling – Fraser River, British Columbia. *Canadian Journal of Civil Engineering*, **17**, 452–459.
Kostaschuk, R. A. and Church, M. A. 1993. Macroturbulence generated by dunes: Fraser River, Canada. *Sedimentary Geology*, **85**, 25–37.
Kostaschuk, R. A. and Luternauer, J. L. 1989. The role of the salt-wedge in sediment resuspension and deposition – Fraser River estuary, Canada. *Journal of Coastal Research*, **5**, 93–101.
Kostaschuk, R. A. and MacDonald, G. M. 1988. Multi-track surveying of large bedforms. *Geo-Marine Letters*, **8**, 57–62.
Kostaschuk, R. A., Church, M. A. and Luternauer, J. L. 1989a. Bedforms, bed-material and bed load transport in a salt-wedge estuary – Fraser River, British Columbia. *Canadian Journal of Earth Sciences*, **26**, 1440–1452.
Kostaschuk, R. A., Luternauer, J. L. and Church, M. A. 1989b. Suspended sediment hysteresis in a salt-wedge estuary, Fraser River, Canada. *Marine Geology*, **87**, 273–285.
Langhorne, D. N. 1982. A study of the dynamics of a marine sand wave. *Sedimentology*, **29**, 571–594.
Ludwick, J. C. 1989. Bed load transport of sand mixtures in estuaries: a review. *Journal of Geophysical Research*, **94**(C10), 14315–14326.

McCave, I. N. 1979. Suspended sediment. In Dyer, K. R. (ed.), *Estuarine Hydrography and Sedimentation*, Cambridge University Press, Cambridge, pp. 131–185.

McLean, D. G. and Church, M. A. 1986. A re-examination of sediment transport observations in the lower Fraser River. Environment Canada, Sediment Survey Section, Report IWD-HQ-WRB-SS-86-6.

McLean, S. R. and Smith, J. D. 1979. Turbulence measurements in the boundary layer over a sand wave field. *Journal of Geophysical Research*, **84**(C12), 7791–7807.

Murphy, P. J. and Aguirre, E. J. 1985. Bed load or suspended load. *Journal of Hydraulic Engineering*, **111**, 93–107.

Rigler, J. K., Collins, M. B. and Williams, S. J. 1981. A high precision digital recording sedimentation tower for sands. *Journal of Sedimentary Petrology*, **51**, 642–644.

Simons, D. B., Richardson, E. V. and Nordin, C. F. Jr 1965. Bed load equation for ripples and dunes. *United States Geological Survey Professional Paper*, **462H**.

Smith, J. D. and McLean, S. R. 1977. Spatially averaged flow over a wavy surface. *Journal of Geophysical Research*, **82**, 1735–1746.

van den Berg, J. H. 1987. Bedform migration and bed load transport in some rivers and tidal environments. *Sedimentology*, **34**, 681–698.

3

Results of Bedload Tracer Experiments in a Mountain River

K.-H. SCHMIDT AND D. GINTZ

Geomorphologisches Laboratorium, Freie Universität Berlin, Germany

ABSTRACT

In the Lainbach catchment, Bavaria, southern Germany, about 1000 artificial handmade concrete and plastic tracers were inserted into the river bed of an experimental reach in different morphological positions (e.g. step, pool, gravel bar) during the years of investigation (1989–1992). These tracers with magnetic cores had defined shapes (ellipsoid, sphere, rod, disc) and weights (100 g, 500 g, 1000 g, 2000 g). They were relocated after individual flood events with magnet detectors. Particles in the pools have the highest probability of entrainment, and these sites are also favoured locations for deposition. The spheres and in some instances the elongated particles cover longer transport distances. The discs of all weight classes, with the exception of the 100 g clasts, show the highest resistance to entrainment and their travel distances are significantly shorter, when small and moderate floods are considered. During a catastrophic flood event in 1990, when the entire bed sediment was in motion, the discs had no transport disadvantage.

The influence of weight on the observed transport lengths is statistically significant only when classes of major differences in weight (e.g. 500 and 2000 g) are compared. Owing to the hiding effect, the small 100 g tracers have much lower maximum transport lengths than the larger particles. The small particles were trapped behind large boulders or in interstices between cobbles and boulders, many of them were deposited in the first few metres of the measuring reach, especially in rough step sections, and were not re-entrained.

Gamma distributions provide the best fit for the observed spatial distributions of the magnetic tracers after the registered events. But the step–pool topography of the longitudinal profile causes irregularities in the distributions. The great influence of steps and pools is demonstrated when the frequency of tracers per metre ($n\ m^{-1}$) is calculated for the individual morphological units. Pools show higher tracer frequencies than adjacent steps.

INTRODUCTION

Transport mechanisms of coarse bedload material under natural conditions are still poorly understood, especially in mountain torrents with a highly variable grain-size mixture, steep gradients and irregular beds. Some of the main fields of interest are the influence of bed topography and particle characteristics (weight and shape) on transport probability and travel length, the pathways of individual particles, the dispersion from point sources and the distribution of distances of movement. Until now very few systematically collected field data sets are available to elucidate this array of problems. In this paper the discussion of some of the above questions relies on results obtained during field investigations with magnetic tracers in the Lainbach, a step-pool mountain torrent in Bavaria, southern Germany.

The Lainbach catchment is located in the Bavarian forealps about 60 km south of Munich. The experimental reach, where the starting points of the tracers were located, lies downstream of the confluence of the Kotlaine (6.2 km^2) and the Schmiedlaine (9.4 km^2) at an altitude of about 750 m (Figure 3.1). Daily mean discharge at the measuring site is about 1 m^3 s^{-1}. During 20 years of discharge

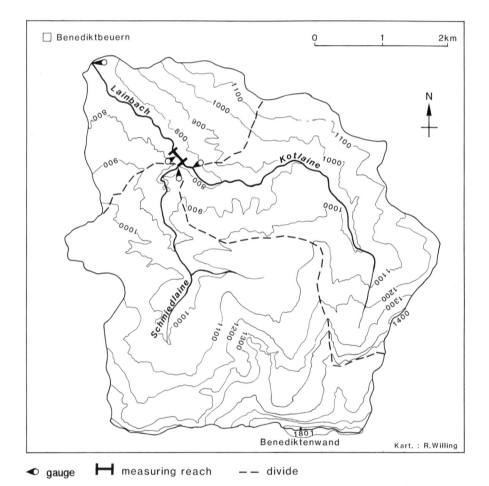

FIGURE 3.1 The Lainbach catchment with the location of the measuring reach (contour interval is 100 metres)

measurements, recorded by the Department of Geography, Munich, no flood exceeded 60 m³ s⁻¹, but in summer 1990 we experienced a flood with an estimated peak discharge of about 165 m³ s⁻¹ with a recurrence interval far in excess of 100 years (Schmidt, 1994). Floods during our summer measuring campaigns usually occurred after long-lasting frontal precipitation and severe local summer thunderstorms. This report refers to moderate floods of the summers 1989, 1991 and 1992.

The Lainbach is a steep mountain river or "Wildbach" in German terminology. Before the catastrophic flood it had an average gradient of 2% in the measuring

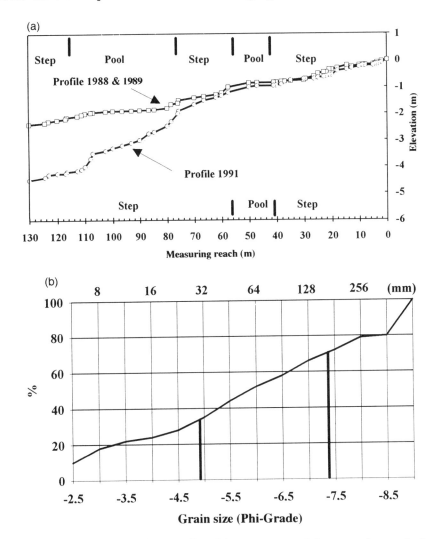

FIGURE 3.2 (a) The longitudinal profile of the upper part of the measuring reach showing step and pool sections before and after the catastrophic 1990 flood. Elevations refer to local datum. (b) Grain-size distribution of river-bed material in a pool section (volumetric sieving of 300 kg of subsurface material including the surface layer). The vertical lines define the section of the cumulative curve represented by the *b*-axes of the magnetic tracers

reach, which increased to about 3.5% after the flood owing to headward erosion behind a breached checkdam downstream of the experimental reach (Busskamp, 1993; Schmidt, 1994) (Figure 3.2a). The long profile of the river bed has the characteristics of a step–pool system. Literature on this type of river has recently been reviewed by Chin (1989) and Grant et al. (1990). Such rivers consist of a series of reaches with steeper gradients (up to 10%) (steps) and intervening reaches with gentler slopes (<2%) (pools). The distinct features of steps and pools are clearly visible in the measuring reach. After the 1990 flood the percentage of the area occupied by steps increased significantly (see Figure 3.2a). There is and there was no regular spacing of steps and pools along the course of the Lainbach.

Bedload sediment is mainly supplied by thick glacial valley fills located in central parts of the catchment. Mass movements and debris flows from the oversteepened valley sides contribute material of widely varying sizes, shapes and angularities to the river system. None of our measured floods was directly affected by mass movements in the measuring reach or upstream of it. There are great differences in the grain-size distributions of steps and pools, with the steps always showing significantly larger mean and maximum sizes (Ergenzinger, 1992; Schmidt and Ergenzinger, 1992). In two pool sections volumetric sieving yielded values of the D_{50} of 50 and 65 mm. Figure 3.2b shows the entire grain-size distribution of a pool sample. In 1991, after the flood, the river-bed roughness had increased substantially and a value of the D_{50} of more than 100 mm was obtained (Busskamp, 1993). The dominant lithologies are limestones derived from the bedrock of the catchment and igneous and metamorphic rocks from the Central Alps found in the glacial deposits. The specific gravities of gravels and cobbles that were measured before the tracer experiments started lay between 2.3 and 2.8. The bed material represents a mixture of shapes, when categorized after the Sneed and Folk (1958) diagram (see Figure 3.4a) with a majority of the clasts belonging to the bladed subgroups. The results of an analysis of about 900 particles are shown in Figure 3.4b. The channels in the step reaches are armoured, the bed surfaces of the pool sections have no firm armouring layer and the gravel bars as well as the stoss and wake accumulations of large boulders are generally loosely structured. After the 1990 flood the inner channel of the lower step in the experimental reach was deeply (2 m) incised into the river-bed surface and became strongly armoured.

FIELD EXPERIMENTS

Some introductory experiments with natural tracers with iron cores were conducted in 1988. The tests demonstrated that there is a statistically highly significant tendency of size-selective transport of the coarse bedload (e.g. $n = 95$, correlation coefficient between weight and travel distance $r = -0.536$, significant at the 0.001 level; for more details see Schmidt et al., 1989), but there is also a considerable amount of scatter in the correlations between weight (size) and travel distance. Only one third of the variance of travel length is explained by the weight and b-axis length of the cobbles owing to the masking influence of other variables, such as the shape of the particles and different starting positions in the river bed (Schmidt et al., 1989; Schmidt and Ergenzinger, 1992). First suggestions of the influence of shape became obvious when travel distances were evaluated in relation to the form categories of Sneed and Folk (1958); a more elongated shape of the particles facilitates longer transport distances, whereas a platy shape has a negative influence on transport length (Gintz and Schmidt, 1991).

About 1000 artificial handmade concrete and plastic tracers with magnetic cores were used to investigate the influence of grain characteristics and river-bed morphology on travel length and the distribution of distances of movements systematically. Plastic was only used for the 100 g tracers. The concrete and plastic tracers lay in the same range of specific gravities (2.4–2.7; mean specific gravity 2.47), the plastic material was mixed with metal powder to adjust the specific gravity. The tracers had defined shapes (ellipsoid, sphere, rod, disc) and weights (100 g, 500 g, 1000 g, 2000 g) (Figure 3.3). The positions of the artificial tracers in the Sneed and Folk diagram are indicated in Figure 3.4a. It must be noted that the 1000 g spheres have a sphericity of only 0.7; the balls of the 500 and 2000 g classes are real spheres with a sphericity of 1 and 0.99 respectively. The geometrical attributes of the different weight and shape categories are listed in Table 3.1.

The particles were inserted into four different positions in the river bed, i.e. a major and a secondary pool (a pool within a cascade reach (according to the terminology of Grant *et al.*, 1990)), a shallow channel in a step, and additionally in gravel bars upstream of large blocks, and a gravel bar near the river bank. The samples in the different positions were given distinct colours for easy identification, each tracer also carried a number on a metal plate (Schmidt and Ergenzinger, 1990). The artificial tracers were placed into their positions by exchanging them with natural stones of similar shape in order to create a near "natural" starting position for first

FIGURE 3.3 The collection of artificial magnetic tracers in the different weight and shape categories

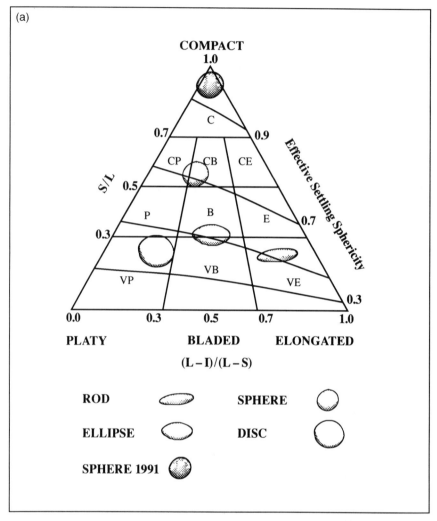
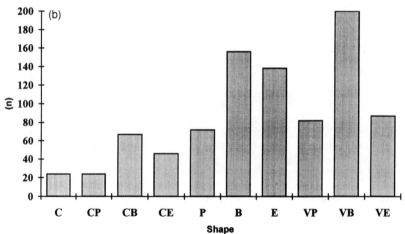

TABLE 3.1 Length of axes and stoss areas of the magnetic tracers

Weight (g)	Shape	a-axis (cm)	b-axis (cm)	c-axis (cm)	Stoss area (cm^2)
2000	Ellipse	19.0	14.0	8.5	127
2000	Sphere	12.0	11.5	11.5	108
2000	Rod	31.0	9.5	8.5	207
2000	Disc	19.0	17.0	6.0	90
1000	Ellipse	15.5	10.0	4.8	60
1000	Sphere	12.0	9.5	6.2	58
1000	Rod	22.0	6.0	6.0	104
1000	Disc	16.0	13.5	3.8	48
500	Ellipse	12.0	8.5	5.0	47
500	Sphere	7.5	7.5	7.5	44
500	Rod	18.5	5.5	5.0	73
500	Disc	13.0	12.0	3.0	31
100	Rod	7.5	3.0	3.0	18
100	Disc	6.0	5.5	2.5	12

entrainment. But it must be borne in mind that the first displacement might be facilitated by non-natural packing and protrusion (Hassan et al., 1991; Busskamp, 1993).

In 1989 the artificial cobbles were manufactured in only one weight class (1000 g), altogether 480 pieces, 120 in each shape category. In 1991 the total collection of tracers was employed including 480 pieces of the 500 g weight class, 120 pieces of the 2000 g weight class, 240 pieces of the 100 g weight class and the remaining clasts of the 1000 g weight class of the previous years. The 2000 g cobbles were inserted only in the pool starting position, and the 100 g sample consisted only of rods and discs made of plastic. In 1992 only those pieces of the original 1991 collection that had been relocated and had remained in "good shape" after transport were employed again. Some of the subsamples had experienced great losses and new tracers had to be produced resulting in a total number of 660 tracers for 1992.

After each flood event the tracers were relocated, removed from their points of deposition and transported back to their starting positions. No cumulative transport distances were registered. Each flood was considered as a separate tracer experiment.

FIGURE 3.4 (a) Shape diagram after Sneed and Folk (1958) with the positions of the artificial tracers. The 1000 g sphere has a sphericity of only 0.7 (position in the CB-segment), whereas the 500 and 2000 g spheres have a sphericity of 1 (position in the C-segment). They were produced for the first time in 1991. Shape categories: C = compact; CP = compact–platy; CB = compact–bladed; CE = compact–elongated; P = platy; B = bladed; E = elongated; VP = very platy; VB = very bladed; VE = very elongated. S = small axis; I = intermediate axis; L = long axis. (b) Histogram showing the shape distribution of natural river-bed material on gravel bars in the Lainbach ($n = 896$)

Relocation of the tracers was started after each of 11 flood events. But only five events in 1989, 1991 and 1992 are used in the following statistical analysis, as in the other cases the search had not progressed far enough before the particle distribution was affected by a subsequent second or third flood event. The dates and respective discharges of the floods that yielded good results are 11 July 1989 (8.1 m^3 s^{-1}), 13 July 1989 (12.2 m^3 s^{-1}), 28 June 1991 (8.7 m^3 s^{-1}), 25 July 1991 (7.4 m^3 s^{-1}) and 22 July 1992 (10.2 m^3 s^{-1}), Between 95% and 71% of the samples were retrieved in the search actions.

RESULTS AND DISCUSSION

Starting positions and sites of deposition

During the field experiments the influence of the starting positions on transport lengths and probabilities was a major concern. Some of the results obtained during the first (1989) measuring campaign with magnetic tracers have previously been reported (Schmidt & Ergenzinger, 1990, 1992; Gintz & Schmidt, 1991) and will not be presented in greater detail. The particles from the pools showed the greatest transport lengths and a 100% chance of being entrained and transported during that year. During the moderate floods only a few of the tracers on the gravel bar near the river bank were displaced. There was no apparent correlation between starting positions and areas of deposition.

The great importance of the pools as bedload source areas was again demonstrated in the following years. The pools are also favoured sites of deposition, which makes them the most active parts of the river bed for bedload exchange and the bedload budget. The steps, on the other hand, with their higher roughness values and steeper gradients, exhibit a relative deficit in deposition. For instance, measurements following the 1992 flood show that 71% of the tracers were deposited in pools and secondary pools with the majority of the cobbles lying in the deeper part of the large pools just upstream of the rise in river-bed elevation. A possible explanation for the surplus of post-flood deposition in the pools can be seen in changing hydraulic conditions in the course of a flood hydrograph. At low stage flow is concentrated in the narrow channels of the steps, it is divergent in the pool sections with a highly reduced probability of erosion. During rising stage the erosion risk in the pools becomes greater. With higher discharge the entrainment probability in the pools apparently increases owing to higher shear forces combined with the reduced effect of grain roughness. At this stage erosion and bedload removal prevails in the pools. In the steps, however, the large boulders, when they are drowned, become part of the grain roughness and flow resistance increases (Grant *et al.*, 1990; Ergenzinger, 1992). Moreover, higher mobilizing forces are needed in the steps due to the structural locking of particles. During falling stage shear forces in the wide pool sections are reduced and they are filled with bedload material, while the intra-flood deposits in the steps are removed by the concentrated flow in the step channels. Some of these changes in bed micro-topography were directly measured with the sounding device "Tausendfüssler" (Ergenzinger and Stüve, 1989; Ergenzinger, 1992). The discussion about the influence of the different morphological units of the river bed

The influence of particle weight

The mean and maximum transport lengths of all weight and shape samples and subsamples are shown in Table 3.2. In general the distributions of the transport distances of the tracers are strongly positively skewed. As the samples are not normally distributed the Kolmogorov-Smirnov test was used to test the data for significant differences in the distributions. When the t-test was applied to the logarithmically transformed data sets to test the travel distances for significant differences of their mean values, the null-hypothesis could be rejected in all cases where the non-parametric test showed significant differences. This also applies to the tests of differences in travel lengths caused by variable shapes. The significance levels shown in the text refer to the Kolmogorov-Smirnov test. The influence of weight on the observed transport lengths of the magnetic tracers is statistically significant only when classes of major differences in weight are compared. The

TABLE 3.2 Mean and maximum travel lengths (m) of moved tracers in relation to shape and weight (g)

Shape	Weight	11.07.1989		14.07.1989		28.06.1991		25.07.1991		22.07.1992	
		mean	max.	mean	max.	mean	max.	mean	max.	mean	max.
Ellipse	2000					307	1186	152	683	464	1576
Sphere	2000					563	1532	326	944	394	642
Rod	2000					290	928	136	518	219	860
Disc	2000					277	950	83	634	135	239
Ellipse	1000	58	164	66	400	460	1468	183	980	237	818
Sphere	1000	53	175	92	345	346	635	197	484	320	640
Rod	1000	59	170	106	430	492	1068	277	403	217	522
Disc	1000	24	71	40	133	266	815	107	530	212	433
Ellipse	500					393	990	275	944	294	930
Sphere	500					566	2000	375	970	414	856
Rod	500					525	2002	261	944	265	854
Disc	500					448	1780	173	897	223	560
Rod	100							200	560	229	679
Disc	100							194	481	191	605
	2000					347	1532	159	944	302	1576
	1000	49	175	74	430	395	1468	180	980	246	818
	500					492	2002	272	970	291	930
	100							197	560	209	679
Ellipse		58	164	66	400	387	1468	203	980	296	1576
Sphere		53	175	92	345	492	2000	299	970	382	856
Rod		59	170	106	430	436	2002	225	944	240	860
Disc		24	71	40	86	330	1780	121	897	201	605

transport lengths of the 500 g tracers are significantly longer than those of the 2000 g tracers in the 1991 floods (0.01 level). In 1992 the distance of the 100 g class is significantly lower than in the 500 g class (0.01 level). In Figure 3.5a and b, characteristic examples for the distributions of travel lengths of the individual weight classes are presented for the two floods in 1991. The skewness of the distributions and the maximum transport lengths (end of the solid line in the whisker

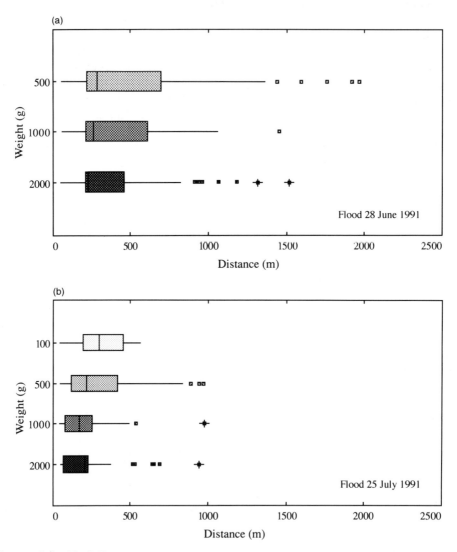

FIGURE 3.5 The influence of weight on travel length for the two floods in 1991, (a) 28 June and (b) 25 July. The central box covers the middle 50% of the data between the lower and upper quartiles, the central line is the median. The lines ("whiskers") extend to the minimum and maximum values up to 1.5 times the interquartile range. When extreme values occur far away from the bulk of the data, they are plotted as separate points (outliers). Very extreme outliers additionally receive a cross

plots) increase with decreasing weight in the 500, 1000 and 2000 g weight classes. This trend is not persistent in the 100 g class. The small 100 g particles have much lower maximum transport lengths, because they are trapped behind large boulders or in the interstices between cobbles and boulders. Many of them are deposited in the first few metres of the measuring reach, especially in rough step sections, and are not re-entrained. It is noteworthy that the favoured areas of deposition for the small 100 g particles are found in secondary pools within the step reaches.

The influence of particle shape

With a few exceptions there have been no investigations on the influence of particle shape on bedload transport parameters (e.g. Komar and Li, 1986; Ashworth and Ferguson, 1989; Gintz and Schmidt, 1991). In the Lainbach experimental reach the influence of shape on travel length and transport probability was for the first time systematically measured in the field. The use of concrete and plastic tracers with defined shapes (see Figure 3.4a) provides a clear data structure for the investigations. The great sample sizes in the shape groups (120 tracers in each shape group in 1989, about 200 tracers in each shape group in 1991 and about 150 in 1992) furnish a sound basis for statistical analysis. Mean and maximum distances for the individual subgroups are listed in Table 3.2. Figures 3.6 a,b and c show the distributions of travel lengths of the different shapes for three flood events.

In the diagram of the flood in 1989 (Figure 3.6a) the spheres consist entirely of the 1000 g weight class, which has a lower sphericity than the spheres of the 500 and 2000 g weight classes (see Figure 3.4a). This might be responsible for the rods having the greatest travel lengths in 1989 (Schmidt and Ergenzinger, 1992). The diagrams indicate that the spheres and the rods cover the greatest median and maximum transport lengths. In most cases, however, there are no statistically significant differences in the distributions of travel lengths of the rods, spheres and ellipsoids. However, their travel distances are significantly higher than those of the discs (0.01 level) (for comparison see also Table 3.2). In the second flood of 1991 and in the 1992 flood the spheres travel significantly (0.01 level) longer distances than the other shape categories. The discs of all weight classes, with the exception of the 100 g clasts, show the highest resistance to entrainment and the shortest mean transport distances, many of them remaining in the starting positions. The example of the flood of 28 June 1991 shows high extreme distances for the discs (Figure 3.6b). Once in motion on a movable bed they can apparently cover long distances. When lying flat on the river bed the discs have the smallest stoss areas (Table 3.1), and they easily rotate into stable imbricated positions. The results demonstrate that the control of particle shape is very significant for coarse bedload transport. The above findings only apply to small and moderate floods. During the catastrophic flood event in 1990, when the entire bed was mobilized, the discs had no entrainment and transport disadvantage (Schmidt, 1994).

The results of the field investigations are supported by experiments with magnetic tracers in the laboratory flume of the Department of Geography in Berlin (Carling et al., 1992; Busskamp,. 1993). Busskamp examined the distribution of shear stresses necessary for the entrainment of the ellipsoid, sphere, rod and disc of the 500 g

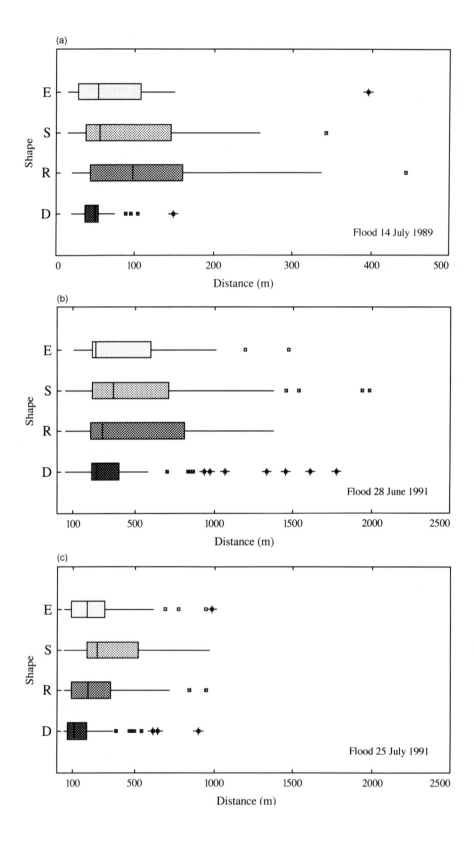

weight class on a rough bed. The tracers were inserted in pockets between roughness elements. He found that the mean shear stresses of the four shapes differ significantly (t-test, 0.05 level), with the sphere needing the lowest mean shear stress values, followed by the rod, the ellipsoid and the disc (Figure 3.7). It was difficult to get enough laboratory values for the disc because it tended to imbricate in the rough bed and could not be displaced even with the highest possible shear stress generated in the flume. Additional experiments (Carling *et al.*, 1992) with varying orientations of the clasts showed that the rod lying perpendicular to the flow direction required the lowest shear stresses for entrainment. In this orientation the rods have the greatest stoss areas (Table 3.2).

When the travel distances during the floods in 1989 and the floods in 1991 and 1992 are compared (Table 3.2; Figure 3.6), the striking differences are obvious. The distances of movement are up to 10 times higher in the years after the catastrophic flood, though the discharges of the floods in 1991 and 1992 were in the same order

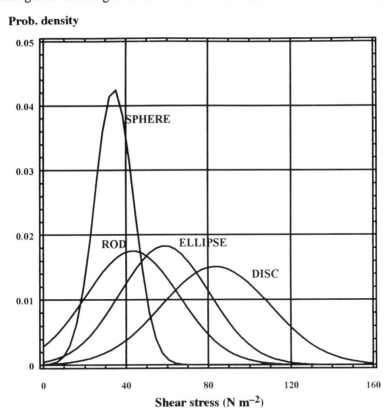

FIGURE 3.7 The influence of particle shape on the distribution of shear stresses necessary for the entrainment of 500 g particles of different shape (measured in a laboratory flume)

FIGURE 3.6 The influence of particle shape on travel length for three floods (a) 14 July 1989, (b) 28 June 1991 and (c) 25 July 1991. The structure of the diagrams is the same as in Figure 3.5

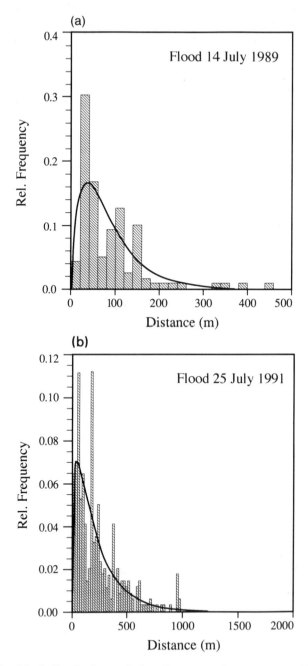

FIGURE 3.8 Empirical distributions of the distances of movement with approximated gamma distributions for two flood events, (a) 14 July 1989 and (b) 25 July 1991

of magnitude. The destruction of a number of check dams in the Lainbach during the 1990 flood has led to an increase in slope in the intervening reaches. This increase in slope resulted in greater available stream power. The higher slope more than compensated for the increased roughness of the river bed. As the total transport length during a flood is the cumulative expression of single steps, it is not surprising that the step lengths of the bedload particles, studied with the radio tracer technique (Busskamp, 1993), also increased significantly.

Distribution of distances of movement

Beginning with H.A. Einstein, a number of authors have discussed the possibilities of a mathematical description of the stochastic distribution of step lengths. Exponential, Poisson and gamma functions were used (Einstein, 1937; Hubbell and Sayre, 1964; Yang and Sayre, 1971; Stelczer, 1981; Schmidt and Ergenzinger, 1992; Busskamp, 1993). The distribution of distances of movements of bedload particles is a question very closely connected with step length distribution as the total transport distance of particles consists of individual steps with intervening rest periods. In recent years gamma functions have been frequently used to describe the distributions of travel distances (Hassan et al., 1991; Kirkby, 1991; Hassan and Church, 1992; Schmidt and Ergenzinger, 1992).

The gamma function provides the best fit for the observed distributions of the magnetic tracer experiments. The empirical distributions are shown for two flood events in 1989 and 1991 with the approximated gamma functions (Figure 3.8). Distances of movement are arranged in 20 m increments. Again the great difference

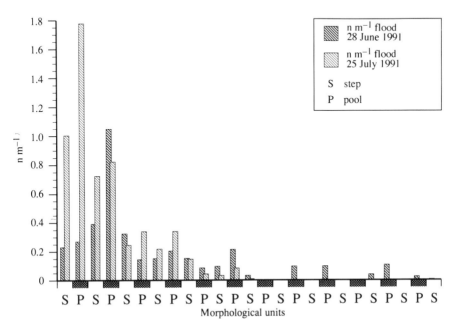

FIGURE 3.9 Frequencies of tracers per unit length ($n\ m^{-1}$) of the major morphological units

in travel distances between the pre- and post-1990 floods becomes visible. There are a number of deviations from the model curve. These are caused by irregularities in the longitudinal profile of the river, demonstrating the important control exerted by the step–pool sequence. The morphologically controlled deviations are responsible for the empirical distribution not being in significant concordance with the model curve. The spatial dominance of the steps and the unequal lengths of the step and pool sections (see Figure 3.2a) complicates the situation. To visualize the influence of the major topographic units of the river profile, the scaled frequencies of tracers per unit length (n m^{-1}) of the major morphological units were calculated. The resulting graph is shown in Figure 3.9. Pools have higher tracer frequencies than adjacent steps. On the whole the number of tracers deposited in pools is 1.6 times greater than in the steps. When the frequencies for the steps are standardized by this simple correction factor, the fit of the gamma distribution can be much improved.

CONCLUSIONS

The magnetic tracer experiments have yielded valuable information on specific problems of bedload research. The measurements corroborate the concept of weight (size)-selective transport for coarser particles during small and moderate floods. There are no strong differences in the transport lengths of the 500 and 1000 g weight classes, but in two of the three floods, in which magnetic tracers of different weight classes were employed, the particles in the 500 g category were transported significantly longer distances than the 2000 g particles. The preceding (1988) iron tracer experiments had shown that very large particles (in the Lainbach 2000 g and larger) travelled shorter distances (Schmidt and Ergenzinger, 1992). This was also observed by Hassan and Church (1992) in other test reaches. The small 100 g particles with b-axes well below the D_{50} of the pool sections in the measuring reach, showed distinctly different travelling attributes. They did not follow the trend of increasing travel distances with decreasing weight and they had the lowest maximum travel lengths (Table 3.2, Figure 3.5b). They were caught in traps between cobbles and boulders especially in rough step reaches. Owing to the hiding effect they were not re-entrained.

Only in this small-sized tracer category was shape of no great importance in controlling travel length. In the coarser particle classes bedload transport was strongly shape-selective. The compact shapes (spheres) and in some cases the elongated shapes (rods) travel the longest mean distances. The platy shapes (discs) of all higher weight classes show the highest resistance to entrainment, travel the shortest mean distances and also, in most cases, the shortest maximum distances, when small and moderate floods are concerned (Table 3.2, Figure 3.6). The small stoss areas and their tendency to imbricate result in shorter travel lengths for discs. During a catastrophic flood event in 1990, when the entire bed sediment was in motion, the discs had no transport disadvantage. For more frequent flood magnitudes the results demonstrate that an exclusion of the influence of particle shape will lead to erroneous results in bedload transport modelling. The low entrainment probability of the discs can be applied in river management strategies to increase bed stability.

In the step–pool system of the Lainbach the tracers in the pool sections had the highest transport probability. The pools are also the most likely locations of deposition. Thus the pools are the most active parts of the river bed for bedload storage and exchange and the bedload budget. Erosion in the pools usually occurs during the rising stage of the floods and deposition during the falling stage. The distribution of the magnetic tracers in the long profile is best approximated by gamma distributions, There are noticeable deviations of the empirical values from the model curve, which are caused by the step–pool sequence of the profile (Figure 3.8). The model convincingly fits the data only for a displacement range where no morphological irregularities control the distribution. When bed morphology exerts influence, the distribution is no longer completely random (see also Hassan and Church, 1992). The significant influence of the step–pool topography is systematically made evident when the frequencies of the tracers per unit length (n m^{-1}) of the major morphological units are calculated. On the average the tracer frequencies in the pools are 1.6 times higher than in the steps. As demonstrated in this paper, particle weight and shape as well as riverbed morphology and structure exercise a substantial influence over coarse particle movement. These parameters must be adequately included in reliable bedload modelling for coarse-grained steepland torrents.

ACKNOWLEDGEMENTS

We thank the Deutsche Forschungsgemeinschaft for the generous financial support of the Lainbach research project. The help of the local authorities (Wasserwirtschaftsamt Weilheim, Forstamt Bad Tölz) is highly appreciated. Our special thanks go to the members of our research group and to the students who eagerly helped during the tedious searches.

REFERENCES

Ashworth, P. J. and Ferguson, R. I. 1989. Size-selective entrainment of bed load in gravel bed streams. *Water Resources Research*, **25**, 627–634.

Busskamp, R. 1993. Erosion, Einzellaufwege und Ruhephasen – Analysen und Modellierungen der stochastischen Parameter des Grobgeschiebetransports. PhD thesis, Fachbereich Geowissenschaften, Freie Universitat Berlin.

Carling, P. A., Glaister, M. S. and Kelsey, A. 1992. Effect of bed roughness, particle shape and orientation on initial motion criteria. In: Hey, R. D., Billi, P., Thorne, C. R. and Tacconi, P. (eds), *Dynamics of Gravel-Bed Rivers*, Wiley, Chichester, pp. 23–38.

Chin, A. 1989. Step pools in stream channels. *Progress in Physical Geography*, **13**, 391–407.

Einstein, H. A. 1937. Der Geschiebetrieb als Wahrscheinlichkeitsproblem. *Mitteilungen der Versuchsanstalt für Wasserbau an der ETH Zürich*, 3–112.

Ergenzinger, P. 1992. Riverbed adjustments in a step–pool system: Lainbach, Upper Bavaria. In: Hey, R. D., Billi, P., Thorne, C. R. and Tacconi, P. (eds), *Dynamics of Gravel-Bed Rivers*, Wiley, Chichester, pp. 415–430.

Ergenzinger, P. and Stüve, P. 1989. Räumliche und zeitliche Variation der Fließwiderstände in einem Wildbach. *Göttinger Geographische Abhandlungen*, **86**, 61–79.

Gintz, D. and Schmidt, K.-H. 1991. Grobgeschiebetransport in einem Gebirgsbach als Funktion von Gerinnebettform und Geschiebemorphometrie. *Zeitschrift für Geomorphologie*, Supplementband, **89**, 63–72.

Grant, G. E., Swanson, F. J. and Wolman, M. G. 1990. Pattern and origin of stepped morphology in high-gradient streams, Western Cascades, Oregon. *Geological Society America Bulletin*, **102**, 340–352.

Hassan, M. A. and Church, M. 1992. The movement of individual grains on the streambed. In: Hey, R.D. Billi, P.., Thorne, C. R. and Tacconi, P. (eds), *Dynamics of Gravel-Bed Rivers*, Wiley, Chichester, pp. 159–173.

Hassan, M. A., Church, M. and Schick, A. P. 1991. Distance of movement of coarse particles in gravel bed streams. *Water Resources Research*, **27**, 503–511.

Hubbell, D. W. and Sayre, W. W. 1964. Sand transport studies with radioactive tracers. *ASCE Journal of the Hydraulics Division*, **90**, 39–68.

Kirkby, M. J. 1991. Sediment travel distance as an experimental and model variable in particulate movement. *Catena Supplement*, **19**, 111–128.

Komar, P. D. and Li, Z. 1986. Pivoting analysis of the selective entrainment of sediments by shape and size with application to gravel threshold. *Sedimentology*, **33**, 425–436.

Schmidt, K.-H. 1994. River channel adjustment and sediment budget in response to a catastrophic flood event (Lainbach catchment, Southern Bavaria). In: Schmidt, K.-H. and Ergenzinger, P. (eds), *Dynamics and Geomorphology of Mountain Rivers, Lecture Notes in Earth Sciences*, **52**, 109–128.

Schmidt, K.-H. and Ergenzinger, P. 1990. Magnettracer und Radiotracer – Die Leistungen neuer Meßsysteme in der fluvialen Dynamik. *Die Geowissenschaften*, **8**, 96–102.

Schmidt, K.-H. and Ergenzinger, P. 1992. Bedload entrainment, travel lengths, step lengths, rest periods studied with passive (iron, magnetic) and active (radio) tracer techniques. *Earth Surface Processes and Landforms*, **17**, 147–165.

Schmidt, K.-H., Bley, D., Busskamp, R. and Gintz, D. 1989. Die Verwendung von Trübungsmessung, Eisentracern mld Radiogeschieben bei der Erfassung des Feststofftransports im Lainbach, Oberbayern. *Göttinger Geographische Abhandlungen*, **86**, 123–135.

Sneed, E. D. and Folk, R. L. 1958. Pebbles in the lower Colorado River, Texas: a study in particle morphogenesis. *Journal of Geology*, **66**, 114–150.

Stelczer, K. 1981. *Bed-Load Transport – Theory and Practice*. Water Resources Publications, Littleton, 295 pp.

Yang, C. T. and Sayre, W. W. 1971. Stochastic model for sand dispersion. *Proceedings ASCE, Journal of Hydraulics Division*, **97**, 265–288.

4

The Interrelations Between Mountain Valley Form and River-Bed Arrangement

C. DE JONG AND P. ERGENZINGER

Institut für Geographische Wissenschaften, Freie Universität Berlin, Germany

ABSTRACT

In valley-restricted mountain rivers, valley form plays an important role in determining the arrangement of sedimentary structures. In this study emphasis is placed on the description of (bed) form roughness, which is considered as part of a hierarchy of roughness features with the final product of system (unit) roughness determining the type of reach formed. In the Lainbach and tributary Schmiedlaine in southern Germany, large- and small-scale sedimentary structures were differentiated according to different types of reaches. In the Schmiedlaine, the river bed can be categorized into a straight bedrock reach subject mainly to erosion in the steepest parts of the valley, into a meandering reach with interactive deposition and erosion, and into a braided–meandering depositional reach. Sedimentary structures are influenced, not only by valley form, but also by associated river-bed and slope gradients. Generally, the lower parts of the valley-side slopes rest at critical angles so that they are easily destabilized by incision of secondary valleys or side erosion. Local inputs of eroded material from debris flows or small tributaries create special river-bed features. Bedrock reaches are mainly sediment conveying and feature large boulders. In the highly meandering reaches, flow concentration occurs before the bend and is followed by a zone of flow expansion and velocity loss below the bend. Large log jams form in the apex of the bends. Levee type "cobble berms" and and/or re-attachment bars develop rings parallel to the outer bend as a result of centrifugal force and related secondary flows. The cobble berms are observed as a hierarchical build-up of complex clusters. Within the expansion zone, the river system becomes subdivided. Different types of bedform systems develop according to their position in the channel or on the bar. On shallow bars at the entrance or exit to the bend, where the reach is relatively straight, diagonal systems of clusters develop, determined by the location of the coarsest material. The same pattern is observed to develop on the shallow bars of the Lainbach. The clusters forming part of the diagonal patterns are of the "imbricate" cluster type. On the proximal parts of the bar, large two-particle clusters are more frequent. In the Schmiedlaine, the distribution of form roughness is strongly dependent on channel curvature. Detailed flow-orientation reconstructions of clusters in comparison to open-bed material reveal that, whereas the open-bed material deviates strongly from the present-day flow orientations and reflects maximum stage flow, the

---

River Geomorphology. Edited by Edward J. Hickin
© 1995 John Wiley & Sons Ltd

coarser cluster material occupies routes parallel to present-day flow in conjunction with the final recessional flood flows. Lobate cross-channel ribs are typically found within less steep channels; i.e. in the lower meandering reaches. In steep channels transverse ribs or step–pool systems develop according to local input and storage of large grain sizes. In both cases extreme events and sediment starvation are a prerequisite for their formation. Although step–pool systems develop independently of the reach geometry their actual location within the Schmiedlaine is channel curvature dependent.

INTRODUCTION

The importance of the spatial variation in bed surface texture in relation to channel form and topography was stressed by Leopold and Wolman (1957). If energy slope and roughness are not consistent from reach to reach, then it is not surprising that the associated bedforms and grain sizes they encompass should vary too (Bridge and Jarvis, 1982). But this detailed differentiation of bedforms is often inadequately described in the literature. Thus, for example, Keller and Melhorn (1978), Richards (1978) and Milne (1982a, 1982b) concentrate on the spacing of riffles, pools and point bars. None of these studies focus on the extremely coarse-grained and steep-gradient streams characteristic of the Alpine streams used in this study. Interactions amongst grain size, channel and valley geometry, longitudinal gradients and flow hydraulics are all indicative of the type of bedform produced. These characteristics need to be linked to the river bed at a reach scale, both longitudinally and cross-sectionally. A clear separation is necessary between the bedform effects on cross-sectional and longitudinal geometry; the long profile has to be considered in terms of long-term river-bed development, whereas the cross-sections often reflect the short-term dynamics (Andrews, 1979).

Form and system roughness (Table 4.1) have been differentiated according to a fractal approach of river-bed analysis (de Jong, 1993). In contrast to grain roughness, form roughness is considered as roughness produced by bedforms whereas system roughness is considered as the largest roughness features that actually influence the type of river system (e.g. step–pool geometry).

Examples of form roughness include the different types of clusters (i.e. imbricate, two-particle, complex clusters, multiple obstacle clusters, transverse ribs, and

TABLE 4.1 Classification of form and system roughness

Form	System
Isolated clast	
Imbricate cluster	Bar
Two-particle cluster	
Complex cluster	Cobble berm
Multiple obstacle cluster	Re-attachment bar
Transverse rib	Step–pool
Megacluster	Bars, log jams

megaclusters), contributing to system roughness (diamond cluster arrangement, cobble berms, re-attachment bars, and normal bars as well as step–pools and log jams). The sedimentary controls include both local inputs of sediment from the slopes and long-term immobile sedimentary deposits from large floods. The stable bedrock curvature induces a cross-sectional symmetry that significantly influences the river-bed morphology. Thus the local bedrock controls the roughness of the river bed more than traditional processes thought to produce rhythmical sequences of steps and pools (Keller and Melhorn, 1978). It has been pointed out by Allen (1982a) that form roughness depends on the size, shape and spacing of bed waves but this direct relation has not been explored much further for natural channels.

In this discussion, reference is made mainly to the Schmiedlaine since it is an example of a natural mountain stream with a wide range of gradients, curvatures and sediments. The Schmiedlaine case study is unique in the following analysis since, in contrast to other cases, the development and spacing of roughness morphology is also dependent on local sediment inputs, dictating grain size. In addition, the main prerequisite for bedform development is an appropriate degree of bend curvature in the confined bedrock meanders.

Data were obtained over a three-year monitoring period, at the beginning of which an extreme flood event with an approximate 150-year recurrence interval occurred (de Jong, 1992a). A detailed survey of a 2 km stretch of the Schmiedlaine enabled the hydraulic geometry and average flood velocity to be reconstructed after the event. The Lainbach provides a basis for comparing examples occurring within the straight reaches of the Schmiedlaine.

The aims of this study are to investigate whether and how the spatial dynamics of form and system (unit) roughness are related to mountain valley form. Special emphasis will be given to valley curvature and how flow orientation can influence the location and type of roughness development. The question of whether small-scale stream patterns can occur independently of channel geometry (Karcz, 1981) is examined. New approaches to roughness description and classification based on recently developed measuring techniques, i.e. photo-sieving and micro-profiling, are introduced. Since not all of the roughness development was monitored during its formation, some of the explanations are speculative but it is hoped that such speculation will stimulate further studies.

STUDY AREAS

The characteristics of the Schmiedlaine and Lainbach study areas are summarized in Table 4.2. In each case, criteria for selection included a heterogeneous, coarse-grained gravel bed, steep gradients and the occurrence of active, sediment-transporting summer floods.

The Schmiedlaine is a tributary of the Lainbach River in Upper Bavaria at the northern edge of the northern Limestone Alps, 70 km south of Munich. Unlike the main Lainbach River, the Schmiedlaine has remained a natural torrent and has not been modified by any checkdams. Details on the catchment, sedimentary and hydraulic characteristics are shown in Table 4.2. The Schmiedlaine constitutes an extremely interesting example of river-bed arrangement and geometry in relation to

TABLE 4.2 Catchment details of Lainbach and Schmiedlaine

Characteristics	Lainbach	Schmiedlaine
Drainage area	15.5 km²	9.4 km²
Drainage length	5.5 km	3.4 km
Max. elevation	1800 m	1800 m
Gradient (average)	0.02	0.05
Drainage density	2.1 km km^{-2}	1.6 km km^{-2}
Annual precipitation (snow)	761 mm	761 mm
Annual precipitation (rain)	729 mm	729 mm
Forest covers	79.1%	75.3%
Width (average Q)	5 m	3 m
Depth (average Q)	0.2 m	0.3 m
Average flood Q (at study site)	20.1 m³ s^{-1}	14.6 m³ s^{-1}
Average Q	1.03 m³ s^{-1}	0.54 m³ s^{-1}
D_{50} (bar surface, study site)	164 mm	86 mm
D_{84} (bar surface, study site)	250 mm	1150 mm

high-curvature, bedrock-confined bends (Figure 4.1a and b). Bed material is very coarse and angular with a large variability in grain-size. The sediment-size data (Table 4.2) are derived from the Lainbach and Schmiedlaine test reaches. Channel gradients (5%) as well as valley slope are steep. In addition, valley-side inputs from small rills, debris flows and slumps form an active part of sediment supply. The forest cover is dense which creates an important organic input in the form of log jams. Vegetation consists mainly of spruce, deciduous and mixed woods.

The geology of the Lainbach and Schmiedlaine catchments is very diverse ranging from Quaternary glacial and glaciolacustrine sediments in the central parts to marls, sandstones, conglomerates, shales and slates in the upper parts (Dobin, 1985; Becht, 1989; de Jong, 1992a; Ergenzinger, 1992). Both valleys were subject to deep post-glacial incision. The investigated stretch of the Schmiedlaine (Figure 4.2) is divided into three main reaches, a straight bedrock reach in the steep, upper parts, a meandering valley confined boulder-bed reach with debris flow inputs in the medial reaches, and finally a meandering-braided, depositional reach. The main measuring site is situated in the last reach, just before the confluence with the Lainbach.

Details on flood hydrology are given in Table 4.2 and in Becht (1989). Large floods occur as a result of short thunderstorm events in July and August produced by the orographic effects of the Benediktenwand, a highly elevated rock face (1800 m). In the Schmiedlaine and Lainbach, the maximum discharge peaked at an estimated 75 m³ s^{-1} and 180 m³ s^{-1} respectively on 30 June 1990.

FIGURE 4.1 (a) Study site in the braided lower reach of the Schmiedlaine, 1992. Notice the high-curvature bend, and inner bend deposition of logs and large boulders. Longitudinal and lateral cross-sections (5–27) are marked in relation to distance downstream. (b) Oblique photograph of same study site. Notice the outer bend berm deposition (light-coloured gravel bars opposite wooden debris in inner bend) equivalent to transects 5–17 on map

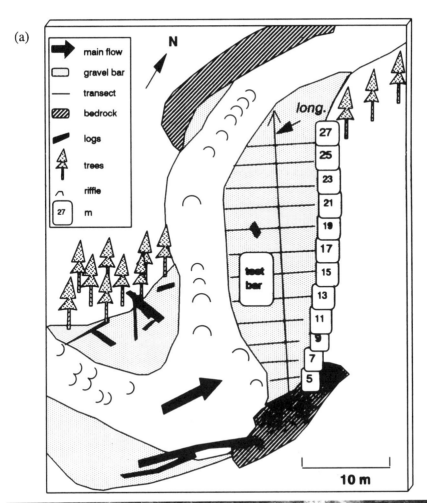

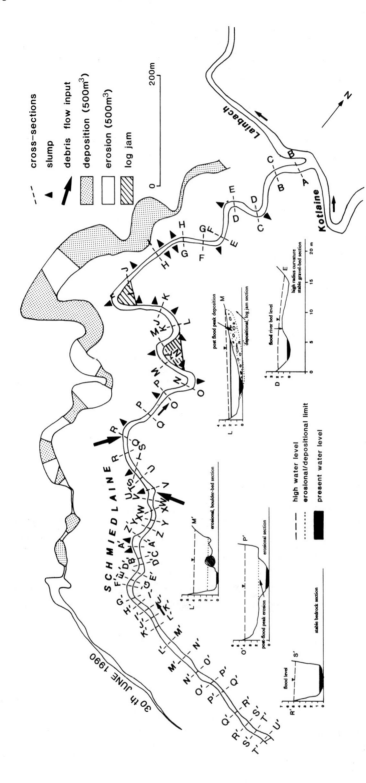

FIGURE 4.2 Detailed geomorphological map of the Schmiedlaine, indicating the three main types of reaches (straight, upper bedrock reach, meandering reach up to IJ and meandering-braided lower reach), the location of the debris flow inputs and resulting log jams. The volumes of sediment erosion and deposition mobilized during the exceptional flood of 30 June 1990 (recurrence interval 150 yrs) are marked as a ribbon above the reach. Corresponding examples of cross-sections marking "typical" reaches with associated maximum water levels are also indicated

The experimental site of the Lainbach lies just below the confluence of the Schmiedlaine and the Kotlaine. Details on the sedimentary and hydraulic characteristics of the catchment are listed in Table 4.2 and in Felix et al., (1988); Ergenzinger and Stüve (1989) and Ergenzinger (1992). The Lainbach valley is broader than the Schmiedlaine and less steep (2% average gradient). It consists of a step–pool system. Unlike the Schmiedlaine, the Lainbach has been modified by an extensive series of checkdams and only at the measuring site is the river bed in its natural state. The area considered in detail is the bar at the measuring site and the general step–pool system along the 150 m long test reach.

METHODOLOGY

Measurements of spatial and temporal changes in roughness and geometry were carried out during the field seasons of 1990, 1991 and 1992 both in the Schmiedlaine and Lainbach. The study is also based on an extensive geodetic survey with 50 cross-sections within a 2 km reach in the Schmiedlaine together with field observation and oblique aerial photography (Figure 4.2).

Problems encountered in the aerial determination of roughness elements have been mentioned frequently in the literature (Sutherland, 1987). In order to determine the role of roughness elements such as clusters, separate analyses for clusters and the surrounding material were carried out. Clusters were identified after Dal Cin (1968), Teisseyre (1977), Brayshaw (1984), Billi (1988), Reid and Frostick (1986) and de Jong (1991) as an assemblage of two or more neatly organized, imbricated particles, limited both in section and in plan to an approximate ellipsoid and forming a "deviation from the general bed level readily detectable to the eye" (Brush, 1965). It is assumed that each cluster consists of an obstacle clast and a stoss or stoss and lee-side accumulation of grais.

Micro-profiling

For the determination of empirical roughness values, the relative projection of particles has to be measured precisely. Micro-profiling of the river bed not only enables the determination of relative roughness, which forms the basis of form drag (Shen et al., 1990), but also the degree of projection, the angle of imbrication, and the shape of individual bedforms in relation to surrounding material. For this purpose a new instrument, the mini-Tausendfüssler device is introduced (de Jong, 1992b, 1993). The mini-Tausendussler micro-profiles roughness by means of vertical needles positioned at 2 cm horizontal intervals. The technique involves setting the micro-profiler parallel to the bed surface and pushing the pins downward until they contact the bed.

Table 4.3 summarizes the frequency and location of longitudinal and cross-sectional transects taken in the Schmiedlaine. Mini-Tausendfussler measurements were limited to the gravel bar due to the hindrance elsewhere of flow depth and velocity and the difficulties of bed surface verification in the water. Profiles were taken in the same direction and along the same profiles as the photo-sieving samples (next section).

TABLE 4.3 Location, direction, number and length of mini-tausendfüssler profiles taken at each study site

Study area	Date	Location	Direction	No. of long-profiles	Total length (m)	No. of cross-profiles	Total length (m)
Schmiedlaine	3 July 1991	gravel bar	long/cross	1	19	15	22
Schmiedlaine	29 July 1992	gravel bar	long/cross	1	33	12	112
Lainbach	4 October 1992	gravel bar	long/cross	2	24	1	12

Geometry

The geometry of the river bed obtained from the mini-Tausendfüssler device was reconstructed at 2 cm intervals for all long- and cross-profiles on the gravel bar and in the channel at the Lainbach and Schmiedlaine. The profiles had to be corrected manually for overlap and were then fitted end-to-end. In addition, all gravel bars were surveyed topographically.

Calculation of particle projection (K_3)

The data obtained from the mini-Tausendfüssler profiles were utilized for calculating the relative roughness or "projection". Projection was calculated from an average value of the moving maximum vertical height difference between three adjacent topographical points as the K_3 coefficient (Figure 4.3; Ergenzinger and Stüve, 1989). Since K_3 was calculated at 2 cm intervals, it is referred to as the K_3 (2). By calculating different K_3 intervals using a fractal approach (Robert and Richards, 1988), these very small intervals allowed the transition between grain, form and system roughness to be determined (de Jong, 1993).

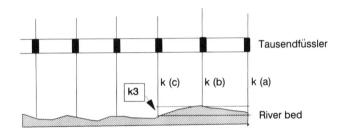

FIGURE 4.3 Calculation of the K_3 coefficient (maximum vertical difference between three adjacent points) as a gliding average value from the Tausendfüssler micro-profiling device

Reconstruction of three-dimensional roughness

From the K_3 (2) data, a three-dimensional roughness model was constructed for the test bar in the Schmiedlaine for 1992. For each mini-Tausendfüssler sample, the maximum K_3 value per sample (1 m in length) was utilized in the plot which consisted of a series of cross-sectional profiles typifying roughness conditions over the entire bar.

Photo-sieving

The method of photo-sieving was first developed by Ibbeken and Schleyer (1986) and extended to particle rounding analysis by Diepenbroek (1992) and Diepenbroek et al. (1992). It was subsequently employed to characterize spatial roughness characteristics in combination with new criteria for particle-size analyses (de Jong, 1992b; Diepenbroek and de Jong, 1993). This method enables complete and undisturbed sampling of the surface layer of the river bed by means of vertical photographs at a given scale (Figure 4.6). Subsequently, the grain size and area covered by a particle, its rounding and orientation can be accurately determined for each single particle digitized and presented in the photographs.

The type and frequency of photo-sieving samples are listed in Table 4.4 and Figure 4.1. Deep flow prevented photo-sieving in the Lainbach test reach. Open-bed (non-clustered) material is compared to clustered material.

Grain-size determination

Grain-size analyses of poorly sorted material common in mountain torrents cannot be obtained in representative quantities through conventional methods. Methods such as those described in Buffington *et al.* (1992) where grain sizes were "determined by randomly lowering a hand-held needle (while the operator looked away) onto the surface and measuring the selected grain-size" clearly require improvement. The photo-sieving method was used to calculate grain-size distributions from the grain b-axes after the Wolman method (Wolman, 1954). Grain size is calculated by transformation of the particle's projected outline into a series of Fourier coefficients. The particle outline is correlated with its best approximating ellipse (assumed to be the ultimate particle's shape after transport) which is obtained from the second harmonic of the Fourier spectrum (Diepenbroek, 1992; Diepenbroek *et al.*, 1992). Photo-sieving provides a much larger sample size per time than other methods used in difficult and extremely coarse-grained terrain. In one hour, the equivalent of 300 kg of sediment can be analysed for all mentioned grain parameters. Although the samples are truncated at 16 mm, the method minimizes sampling error and ensures a very fine resolution for the coarser dominating size classes.

Comparisons of the grain-size information were made with the K_3 roughness data in order to validate the new coefficient. The D_{16}, D_{50} and D_{84} were chosen as conventional indicators of roughness. In all comparisons of grain size and K_3 roughness, the K_3 (2) (the smallest possible interval) was used in order to represent all minor roughness variations.

Grain orientation

Detailed grain orientation studies have also been neglected due to the intensive labour involved in manual-orientation measurements in the field. The photo-sieving techniques provide the opportunity for rapidly and accurately calculating the horizontal orientation of individual particles according to the grain a-(long) axis (Figure 4.5 and Diepenbroek, 1992; Diepenbroek *et al.*, 1992; Diepenbroek and de Jong, 1993).

Orientation data were processed and plotted as rose diagrams. At each cluster site, the orientation of each cluster was calculated separately from the average orientations of the surrounding open-bed material. The orientation of a cluster was

TABLE 4.4 Location, direction, total sample area and size of the photo-sieving sample together with actual sample area and size processed (open-bed) as well as sample area and size of clusters and cluster particles. Cluster density is expressed as the area one cluster occupies per square metre

Study area	Date	Location	Direction	Total sample area (m^2)	*Open-bed area (m^2)	Flow orient. sample	*Open-bed grain sample	Cluster particles (no.)	Cluster bedforms (no.)	Cluster particle area (m^2)	Cluster density (m^2/m^2)
Schmiedlaine (pre/post-flood)	7 July 1990	test bar	long	37.80	30	2000	1374	191	17	2.95	1.61
	18 July 1990	test bar	long	45.36							
Schmiedlaine	3 July 1991	test bar	long/cross	60.48	15	5000	1339	54	12	3	0.56
Schmiedlaine	30 July 1992	test bar	long/cross	64.26							

* Sample processed from total (clusters not included)

obtained from the whole cluster. Results were combined and plotted on a geomorphologic orientation map (Diepenbroek and de Jong, 1993).

Gradient

The relation between channel gradient and boundary roughness is mentioned briefly in the literature (Osterkamp, 1978; Wilberg and Smith, 1987; Ergenzinger and Stüve, 1989; Grant *et al.*, 1990; de Jong, 1992a). The largest types of bedforms are expected in high-gradient river reaches, such as steps or steep bedrock canyons. Gradients in this study were recorded with a theodolite in 1990 and with an infrared laser in 1991–1993 from more than 50 cross-profiles along a 2 km section of the Schmiedlaine. In addition, local gradients along the gravel bars were extracted from mini-Tausendfüssler long profiles.

Curvature

The degree of river curvature and the amplitude of the bend play a decisive role in the distribution and size of roughness. The radius of curvature was obtained from river sections along a detailed geomorphological map, reproduced from 50 cross-sections.

RESULTS AND DISCUSSION

Form roughness

The form roughness described in the following sections has to be considered within the hierarchy of bedforms (Allen, 1968) presented in Table 4.1. For example, clusters and transverse ribs are superimposed on the bar, transverse ribs in the channel, complex clusters form part of cobble berms, and megaclusters are components of log jams. Form roughness has been separated into individual classes of clusters/transverse ribs since they are an important reflection of individual flow and sediment transport "streets", i.e. the accumulation of bedforms into system (unit) roughness does not grow homogeneously (Bagnold, 1956).

Imbricate clusters

These are particle assemblages (Laronne and Carson, 1976; Brayshaw *et al.*, 1983; Brayshaw, 1985; Billi, 1988; Hassan and Reid, 1990; Reid *et al.*, 1992) on average three to six particles in length and form as single, highly imbricated longitudinal threads. They need not consist of the largest material on the bar, even though this is predominantly the case. They are differentiated in terms of shape, number and arrangement of particles. These types of clusters are very frequent in all types of environments in both the Schmiedlaine and Lainbach (Figure 4.4). In steep reaches such as in the Schmiedlaine, their angle of stacking may be vertical (Figure 4.4a). In such cases particles will be deposited behind oversized obstacle clasts. Such clusters will be found as separate entities on the bar top (Figures 4.4a and b) or along the bar

(a)

(b)

FIGURE 4.4 Examples of typical imbricate clusters in the Schmiedlaine (a) in a dry channel on the bar, reach above EF in 1990, flow from right to left, (b) in a dry channel on the bar, reach above IJ in 1992, flow from right to left, (c) in section at channel edge, at DE, 1992, flow from left to right and (d) the same cluster as (c) in plan

(c)

(d)

FIGURE 4.4 (*continued*)

edge (Hassan and Reid, 1990, and Figure 4.4c and d). In the Schmiedlaine, they are also located in medial positions of the secondary channels that are active during high flows and restricted to the middle to lower reaches of the stream. This is true for small as well as mega-sized clusters. They are well adapted to flow, their ellipsoidal shapes in plan and in section form a roughness entity that offers less resistance to flow than an *ad hoc* assemblage of single, open-bed clasts.

In the Schmiedlaine, as in the Lainbach, it was observed that freshly deposited bar areas recently uncovered after the occurrence of a flood were densely covered with imbricate clusters. This was especially the case for the proximal bar regions. Since this upper part of the bar is also usually the coarsest, this means that there will be numerous particles eligible for cluster formation. Clusters constitute the coarsest fraction of the river bed. From detailed flow-orientation studies on the test bar it is clear that coarse bedforms are deposited along the low-energy routes during the waning flow stages. Photographic documentation and observations show that entire bar surfaces are deposited during the high-flow stage, whereas a coarse lag of material composed of clusters and similar bedforms is deposited towards the final stage of the flood.

The influence of flow on the spatial variability of roughness

The orientation of clusters was compared to the mean orientation of all surrounding material at each cluster site in order to establish former flow conditions over the new bar. Cluster orientations were obtained from the actual outline of the cluster since the variation obtained from the single particles was often too high.

The ability of flowing water to get particles into motion is a function of particle properties and their arrangement on the river bed (Sutherland, 1967). In these coarse gravel-bed streams it is assumed that the particles are aligned parallel to flow (Allen, 1982b). But depending on the grain shape and position within a cluster, not all particles will assume this direction. Many particles will rest in orientations that give them the best protection against disturbance, which explains why particle arrangement is so important in incipient motion (Klingeman and Matin, 1993). A cluster may consist of an obstacle lying transverse to flow, while the remaining stoss-side accumulations will be oriented parallel to flow. It is for this reason that the entire cluster shape was utilized in the flow orientation study.

Flow orientations of open-bed particles did not deviate by more than 30° on average. As is evident from Figure 4.5, two very different flow histories can be deduced from the difference in orientation between clusters and open-bed particles for the flood flows of 1990. Whereas clusters generally reflect straighter routes parallel to the present-day flow, the open-bed particles tend to diverge consistently from the clusters to a greater or lesser extent depending on their location in relation to the bend. This divergence is not a measurement error but an indicator of flow reworking clusters in relation to surrounding material.

The consistent flow arrows in Figure 4.5 show that open-bed clasts are subject to flows that cut straight across the bar and channel at the proximal bar end downstream of the wooden debris. These are then violently deflected off the bedrock valley slopes in the opposite direction, only to be straightened again at the distal bar end. This

70 *River Geomorphology*

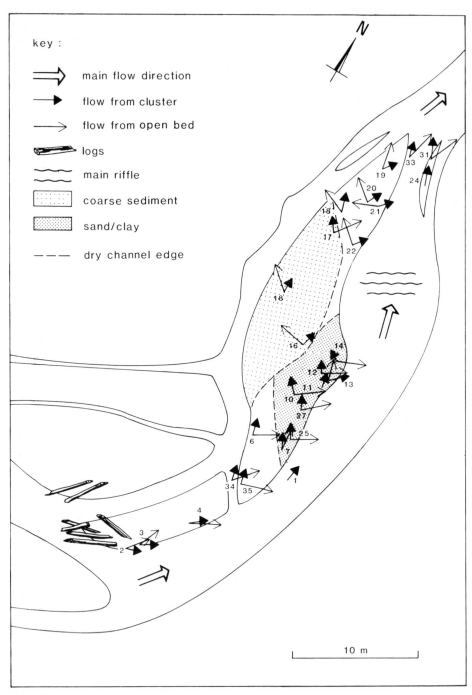

FIGURE 4.5 Comparison of mean flow directions from orientation of clusters and open-bed material on the test bar in the Schmiedlaine for 1990. Numbers indicate cluster sample locations. Note large divergence in flow orientation over gravel bar and more similarity at entrance and exit to bent

lateral deflection amounts to 100° in some cases. The open-bed material flow orientations are very much dependent on bend curvature. At the beginning of the bend, open-bed sediments are fairly parallel to the clusters, but a deviation in grain orientation occurs at the proximal bar end, with open-bed material pointing straight across the bar. As a result of centrifugal forces, open-bed material is forced around the bend and is subject to deflection in the opposite direction. The renewed straightening further downstream is the result of flow convergence and concentration at the exit of the curve.

Thus, open-bed arrows reflect flood hydraulics. In Figure 4.5 the character of the straight flow directions upstream of cluster 14 shows that high flood stage must have existed in order for the equally oriented particles to have been deposited over the entire bar top. The equivalent pattern of flow occurring at the distal bar end is supported by a number of well-spread sample sites all indicating fully developed flood flow.

In contrast, the clusters, which are coarser in size than the open-bed material (de Jong, 1993), indicate that flow directions are well adapted to the individual geometries of the major and minor bars (Figure 4.5) and are more similar to present flow conditions. Brayshaw (1984) also found that clusters exhibited more consistent alignments parallel to flow. It is therefore suggested that clusters reflect the final declining and depositional stages of the flood.

The data processed for the floods of 1991 indicate a similar pattern of differentiation between clusters and open-bed particles, except that the range of deviation between the two is less (only 46°). Since the clusters did not deviate as much from the open-bed material (i.e. clusters were oriented in directions very similar to the open-bed and present flow directions), it is postulated that, for 1991, the floods succeeding the major 1990 events were less extensive and powerful.

Two-particle clusters

These occur very frequently on the finer-grained river bed, mostly in conjunction with larger particles (Figure 4.6). They are found on most gravel bars and could possibly be immature clusters. It is noteworthy that the stoss-side clast behind the very rounded obstacle usually covers the same area but is far more platy, thus decreasing the chance for any smaller particles to be deposited behind it. At Lainbach, these forms are predominant on the proximal bar tops. In the Schmiedlaine they are restricted to the low-gradient reaches and are less frequent because the wide variety of angular material increases the chances of longer clusters being formed instead. It can be seen in Figure 4.6 that the shape of the overall cluster is optimally adapted to flow by offering least friction. An odd-shaped or oversized stoss-side particle would most probably be subject to lift forces (Ergenzinger and Jupner, 1992) and would be entrained by the flow. It is therefore important for the second particle to be well interlocked with the obstacle clast.

Diamond arrangement of clusters

This term has been given to the distinct diamond pattern that imbricate and two-obstacle clusters adopt primarily on the bar within a straight reach (Figure 4.7). This

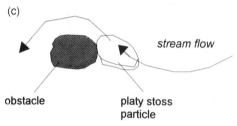

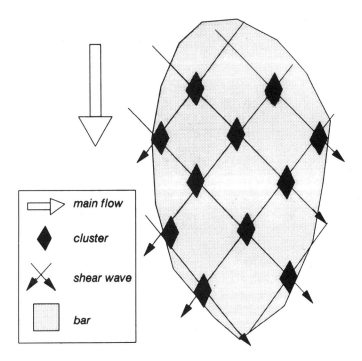

FIGURE 4.7 A model of shear wave development over a gravel bar. Clusters form where shear waves intersect. Longitudinal roughness measurements during floods show that shear waves are necessary for the deposition of cluster obstacles

type of arrangement is found on practically all gravel bars that are covered by shallow flows during floods (Figure 4.13). Thus, in the Schmiedlaine, clusters are never arranged directly downstream of one another. Instead, they are located at the right and left downstream extremities of the cluster. In the zone below the cluster the wake-zone turbulence probably inhibits the direct deposition of another bedform. Diamond clusters can take on various shapes, depending on the bar gradient and shape of material. As is suggested by Hassan and Reid (1990) and explained below, their spacing is an equilibrium characteristic.

Regular lattice-like arrangement of bedforms occur when transverse and longitudinal rows of particles join into rings. Similar ring-like diamond patterns of clusters have been produced independently in coarse-grained laboratory experiments (Allen, 1982a; Tait and Willetts, 1992). Each rhomboid forms like an elongated diamond (Tait, pers. comm.; Allen, 1982b). Other authors have termed them stone cells (McDonald and Banjeree, 1971; Gustavson, 1974) which they related to shallow-flow surfaces.

FIGURE 4.6 Two-particle cluster on bar in Lainbach. Flow is from right to left. (a) In section, (b) in plan and (c) schematically. Note that the stoss particle behind the large rounded obstacle in (a) has a flatter side protruding to flow (as in Figure 10)

Allen indicates that rhomboidal surface waves (or shear waves) are responsible for their formation at Froude numbers above 1, although he restricts his observations to sand beds and muddy beaches. Karcz (1981) explains how obstacle-induced oblique hydraulic jumps cause the formation of rhomboidal patterns, where wave trains are refracted by the obstacle across the stream path. Centrifugal components distort the hydrostatic pressure distribution three-dimensionally so that a secondary motion causes the longitudinal acceleration of fluid and the production of the diamond pattern. Although descriptions are lacking for gravel beds, the rhomboidal lattice structure of clusters may be related to the regular criss-crossing streamlines formed both during the descending flood limb in the Lainbach and in other generally low-gradient bar tops in the low-energy zones between the bends of the Schmiedlaine (Figure 4.7). Their tendency to form in shallow depths, on gently sloping bar surfaces with smaller grain sizes, is evident from similar laboratory experiments by Tait (pers. comm.). In the Schmiedlaine, clusters were mainly of the imbricate type arranged as separate entities (i.e. in an unjoined diamond pattern). From photographs of the shallow flow depth (<10 cm) during the receding flood and from longitudinal measurements of the corresponding re-building of roughness elements, it is clear that shear waves are necessary for the deposition of cluster obstacles under the given subcritical flow conditions.

The diagonal separation of flow lines can vary according to the obstacle size and shape. Approximately one obstacle radius either side of the centre-line, the vector field is convergent or depositional. This phenomenon is essential in determining the location of the next cluster. The obstacle shape has considerable influence on the formation of roughness patterns and irregularities in nature (de Jong, 1993). Thus, larger obstacles attract larger clusters in the proximal parts of the bar at the end of the bend, mostly of the two-particle type. As the bends straighten out, clusters become longer and consist of a greater number of smaller particles.

Hodges (1982) has described shear waves or rhomboidal flow on slopes. These shear waves are wave patterns flowing diagonally towards one another and crossing each other at nodal points such that a diamond pattern is induced. At the intersection point, the flow velocities are reduced thereby enabling an obstacle to become stranded. Once it is stranded, it determines the location of the next shear wave intersection point necessary for subsequent obstacle deposition. In contrast to Allen (1982a), he attributes them to the flow recessional stage, as discussed here, where Froude numbers (Fr) are subcritical. Temporal measurements of roughness change in relation to flow velocity and river geometry on the bar in the Lainbach also indicate that $Fr < 1.0$ for the formation of shear waves. The braided structure of flow is limited to low-gradient surfaces, as on bars.

Shear waves may wander according to the movement of isolated particles. Comparisons with flume experiments under steady flow (Lawrence, 1987) showed that when a hydraulic jump exists over an obstacle and an obstruction is put into the flow above the obstacle, the hydraulic jump will move upstream of the obstacle and can remain there even if the obstruction is removed. This shows that a temporary flow disturbance will suffice to produce a longer-lasting shear wave effect even in the absence of obstacles.

Complex clusters

In contrast to the imbricate clusters these assemblages are quite chaotic in arrangement and can be likened to a proto-bar form that is dome-shaped in cross-section. They protrude higher into the flow than imbricate clusters and are also more than one particle in width. In many cases they have the same number of particles in length as in width (5:5, see cross-section, Figure 4.11b) and usually consist of equally sized particles. In general they may also be oval in shape and consist of the coarsest clasts on the bar, thereby protruding above the general bar level. Sometimes they consist of an outer wall of coarser clasts with an infill of finer material. The coarse cobble berms in the Schmiedlaine consist almost solely of an assemblage of these clusters. These are located primarily in the sharp river bends where channel width is not restricted, i.e. in the middle to lower reaches. Specific locations include MO, KL, EF and CD (Figure 4.2). They require high gradients ($\approx 4\%$) and are located exclusively in the outer regions of high-curvature bends.

Multiple obstacle clusters

These anabranching, fork-shaped clusters consist of two to three rows of pebbles each locked against an obstacle clast and growing upstream towards a mutual assemblage of particles (Figures 4.8a, b and c). The cluster in Figure 4.8 is located mid-channel in the lower part of the reach in the outer bend just after the bend apex (Figure 4.13). These types of clusters were found during all years in the Schmiedlaine only. It is not yet clear whether they form in an upstream or downstream direction. As observed by Karcz (1968) in sandy media, a single obstacle can actually cause a series of ridges to form downstream of it.

The reason for their formation in specific channel locations is that they require very coarse material (>20 cm *b*-axis) and steep lateral bar gradients. Multiple obstacle clusters are mostly found in the active part of the channel (i.e. a further development of imbricate clusters), in the erosional part of the reach. The angle between two obstacle-induced rows of particles may attain more than 90°. Thus at the test bar in the Schmiedlaine, they were found above the depositional cobble berms just at the beginning of the bend towards its outer edge. In the reach below the debris flow input, at PQ (Figure 4.2), a two-obstacle cluster was also located in the active, erosional part of the channel. The very wide angle of division may characterize the initial stages of arcuate transverse ridge formation. Such dynamically formed bedforms are therefore restricted to high-energy zones or zones of flow concentration at the entrance and exit to bends.

Transverse ribs

Transverse ribs (McDonald and Banjeree, 1971; Gustavson, 1974; Boothroyd and Ashley, 1975; Koster, 1978; Naden and Brayshaw, 1987) consist of lateral ribs of clusters. They are formed during the waning stages of flow in shallow, high-energy environments indicating that they require high velocities and a high capacity of sediment transport. Transverse ribs are characteristic of low-sinuosity single-channel

(a)

(b)

FIGURE 4.8 Multiple obstacle cluster in the Schmiedlaine, (a) photograph in plan, 1991, upper test bar, flow from left to right, (b) oblique photograph, reach PQ, flow from left to right, and (c) sketch (in plan) of cluster in (b), flow from right to left. Numbers are the hypothesized order of stoss particle arrangement after the deposition of obstacle clasts

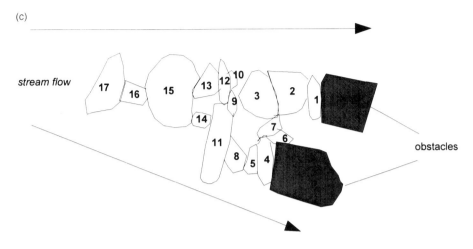

FIGURE 4.8 (*continued*)

reaches with high width to depth ratios. In these cases, with the exception of work by Laronne and Carson (1976), transverse ribs are linear and are restricted to glacier outwash plains where coarse ribs are found in intervening areas of silt.

In the Schmiedlaine, they are composed entirely of coarse projecting sediment and take on four separate forms (Figure 4.9): a transverse rib or minor step located perpendicularly across the entire channel width, a rib located diagonally across the channel, an arcuate rib located across the channel, and an arcuate rib interrupted in mid-channel location. These bedforms are typically found at the exit to channel bends where the grain sizes fall in the cobble to gravel range and flow is convergent. Transverse ribs preferably form in the depositional reaches; they cannot be detected in the erosive reach. It is suggested that, in steep and narrow channels or in the vicinity of very coarse material, transverse ribs are replaced by up-scaled step–pool formations.

These transverse steps are formed in areas of flow acceleration at the entrance and exit of bends, where the higher gradients allow standing waves to form. Comparisons with studies in nature (McDonald and Banjeree, 1971) and in the flume (Grant *et al.*, 1990) indicate a local congestion of material underneath standing waves may be a possible reason for their formation. Once such a step is formed, a local standing wave develops over it and it is the amplitude of the wave that probably dictates where and in which dimensions the next step will be formed. They form under supercritical conditions, (Koster, 1978) typically on the bar tops.

Once a full standing wave is developed across the channel during the receding flood, a mature transverse rib will project across the entire width of the channel (Figure 4.9). In a favoured channel reach, bedforms can be found in close proximity to each other in various stages of development, including the "destroyed" stage (Figure 4.9a). This type of incomplete transverse arc has been dissected in mid-channel location, most probably as a result of the associated higher velocities during flows superseding the event that formed them. In pool regions between these steps

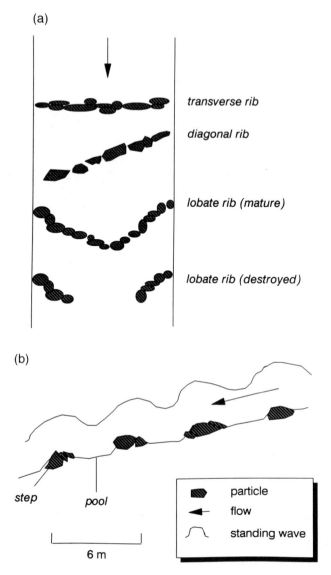

FIGURE 4.9 Pattern of formation of transverse ribs in the Schmiedlaine (a) in plan form, transverse or diagonal to flow and (b) in section, with standing waves over steps, and troughs over adjacent pools

flow is fully concentrated, velocities are high causing scour underneath the wave trough. The next transverse rib will thus be deposited under the succeeding standing wave. Depending on the channel configuration and planform, different types of transverse ribs may be formed. Where the channel is starting to curve into the bend, transverse ribs may extend diagonally across the channel (Figure 4.9). In other reaches, where the channel is incised more deeply, the rib does not form straight across the channel but is deposited in a semi-circle pointing downstream.

Megaclusters

Megaclusters are defined as outsized clusters on bar tops (Figure 4.10). They are so large that they can form part of system roughness in the upper reaches of the Schmiedlaine. Megaclusters are dependent on local sediment supply inputs from slopes, e.g. debris flows. The particles constituting megaclusters can be transported only for short distances. Megaclusters are depositional features and they are often found as singular exceptions towards the end of the bar. On the Schmiedlaine test bar, at CD (Figure 4.2), the single very large cluster at the distal end of the bar, just prior to the entrance of the next bend, also formed the roughest feature on the bar.

In the reach below the influence of the 1990 debris input, at PQ (Figure 4.2), the best example of a megacluster (3 m in length; see Figure 4.10) is in a large dry channel on the medial bar. The bar has an average gradient of approximately 5° and in this case the cluster has a major influence on flow divergence. In terms of flow direction, these types of mega-forms reflect the most direct flow routes. In this case the obstacle clast is again the main determinant of the shape and size of cluster formed. The obstacle probably induces supercritical flow (Zgheib, 1990) which through vertical and horizontal energy dissipation dictates the further development of the bedform. Other types of megaclusters in the very steep middle to upper reaches of the Schmiedlaine show how prone large clasts are to location-specific accumulation. Clusters, controlled by the role of the obstacle, therefore play an important function in river-bed sorting and local increases in roughness.

FIGURE 4.10 Photograph of megacluster in the Schmiedlaine, reach PQ. Flow is from right to left. Cluster is 3 m in length. Notice the rounded obstacle clast, tight particle interlocking and infill, and the flat surface of the final stoss particle

System roughness

Cobble berms

These features are the result of an accumulation of complex clusters into arch-shaped, levee-like structures (Figure 4.11a and b). Their formation can be likened to those of boulder berms (Carling, 1989). Flume experiments suggest that they form

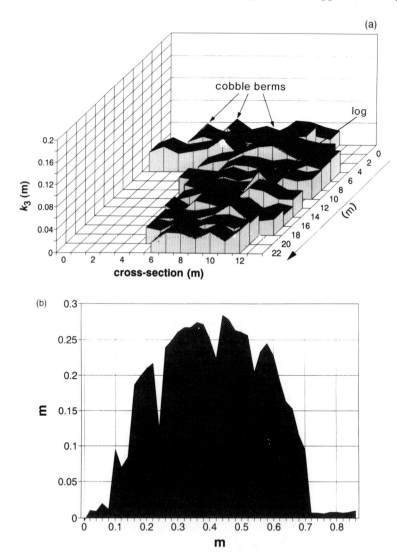

FIGURE 4.11 (a) Three-dimensional model of roughness development calculated from average K_3 values along transects on the lower test bar of the Schmiedlaine, 1992. Flow is towards the observer. Note the location of cobble berms in the outer bend. (b) Single mini-Tausendfüssler cross-profile of a complex cluster, constituting the cobble berms, in 1991. Note optimal arrangement of five separate particles into flow-dynamic shape

under unsteady supercritical conditions and are rapidly deposited independent of sediment transport rates. Their occurrence is frequent in the very steep, high-curvature bends of the Schmiedlaine (Figures 4.11 and 4.12), such as at the test bar, reach CD, at NO and NM (Figure 4.2).

From evidence gathered in the Schmiedlaine during several flood events over the three year period, it is obvious that they form as a result of sequential outer channel deposition of arch-shaped levees (Figure 4.12) within a single flood. It is assumed that these features mark a recessional stage of river-bed formation. The deposition of these arcuate bedforms is dependent on the radius of the flow in the bend which in turn is a function of the discharge (Figure 4.11). At high discharge, photographs show that flow will be fully developed across the channel, but, as it recedes, the first cobble berm will be deposited at the very outer edge of the channel. As flow recedes further, the next cobble berm in the series will be deposited adjacent to the first. This process continues three or four times until flow has retreated to its former channel limits. From detailed grain-size analyses on the Schmiedlaine test bar and from photographic observations further upstream, it is evident that there is a size gradation from the outer towards the inner bend. This is not surprising since the coarsest sediment will be deposited under high discharge in the outer bend and finer material will be deposited subsequently as the energy gradient declines with falling discharge.

Due to the shape of the channel in the Schmiedlaine it is not possible for these features to form by mid-channel flow separation. According to Carling (1989), the deposition of the berms is the result of a minimization of flow energy in the outer bends during high flow although he excludes the role of sediment transport in their formation. Since these cobble berms are stage and discharge dependent, they, however, must depend on sediment transport as well. The gradual deposition of sequences of berms away from the outer bend into the main stream indicates that not only discharge but also the sediment transporting capacity must be decreasing rapidly.

In the three-dimensional model of the K_3 roughness distribution on the test bar in the Schmiedlaine (Figure 4.11), two parallel high roughness areas associated with cobble berm deposition are indicated in the left background. The remaining roughness areas peak at regular intervals of 6 m associated with the deposition of major roughness elements underneath standing waves developed during flood flows. The curvature of the channel is therefore decisive in trapping coarse sediment in its outer bend, and distributing roughness at regular intervals thereafter.

Re-attachment bar

The only extensive description of a re-attachment bar documented so far originates from the work by Rubin *et al.* (1990). They are narrow, elongated depositional features that are located within the channel expansion zone where a weak recirculation current developed. Although described in a larger scale study treating finer grain sizes, their formation in the outer bend where flow is beginning to expand can also be applied to the Schmiedlaine. This type of feature presents the same scale as the cobble berms. As in the case of cobble berms, re-attachment bars consist entirely of complex and imbricate clusters. They are different in that they form in

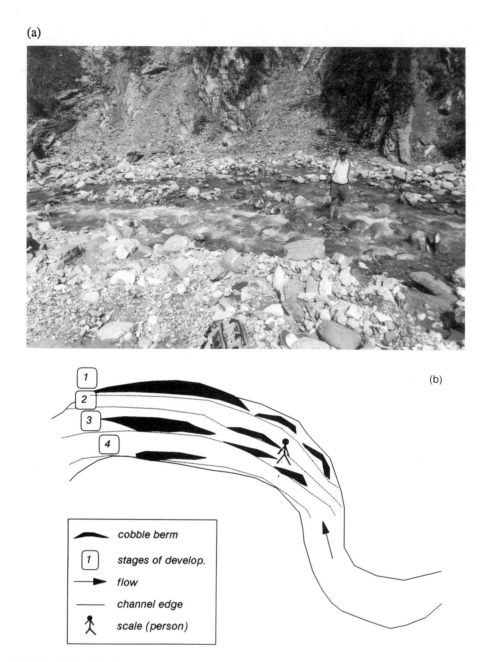

FIGURE 4.12 (a) Photograph of sequential outer bend deposition of cobble berms in high-curvature reach NO. (b) Sketch of hypothetical cobble berm formation, numbered in order of formation (1–4). Cross-channel gradient is declining towards observer

very narrow bends and are attached by one end to a bedrock spur. Occurrence of this type of feature obviously is dependent on valley curvature. Where the radius of curvature is very small, as in the case at cross-section DE, the depositional zone is so compressed that the re-attachment bar is forced to adopt a narrow elongated shape.

Step–pools

Step–pool systems are found in steep, coarse-grained environments (Whittaker and Jaeggi, 1982; Chin, 1989; Ergenzinger and Schmidt, 1990; Grant et al., 1990; Ergenzinger, 1992). Few studies exist on the patterns of step–pools and these are not directly comparable to the more widespread riffle-pool sequences due to the difference in grain size. In the Schmiedlaine, step and pool units are mostly situated in the straight reaches or at the entrance and exit to the bend, not in the bend apices (Figure 4.13). They are found throughout the upper and middle reaches where grain sizes are large and the channel is steep. In the Schmiedlaine and Lainbach, pools are much finer-grained (Ergenzinger and Stüve, 1989) than steps. Exceptionally large or long steps generally exist in the higher gradient channel bends where there is a higher probability of coarse-sediment deposition or direct slope inputs. Steps are asymmetrical in cross-profile, whilst pools are generally more symmetrical and developed in the straighter reaches. This stands in contrast to the riffle-pool patterns observed by Keller and Melhorn (1978) and Milne (1982a).

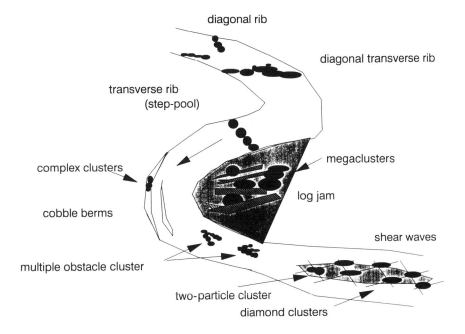

FIGURE 4.13 Summary of the role that the bend plays in the Schmiedlaine in determining form and system (unit) roughness development. Note formation of roughness features typical of the broad and narrow straight reaches, the entrance and exit to the bend and the inner and outer bend

The formation of step–pools (Grant et al., 1990) has been attributed to the development of standing waves and supercritical flow, a process scaled up from transverse ribs. The standing wave that develops over a step will cause erosion downstream and the associated development of a pool (Figure 4.9).

For a number of reasons, a regular step–pool pattern is not well developed in the Schmiedlaine. Although step and pool units repeat at regular intervals, the actual distance from one step–pool sequence to the next remains highly variable and dependent on channel curvature. Step–pools are also dependent on grain size within their local environment, influenced by slope inputs and longer-term immobile flood deposits (Grant et al., 1990). In addition, the influence of local geology interrupts regular patterns. Where a local input of very coarse material is supplied to the channel, the river is incapable of transporting it away and steps are initiated. In other cases the steps are composed of bedrock. Since the grain-size distribution is highly variable downstream (Milne, 1982b), certain locations will favour large differences in gradient and the deposition of coarse grain sizes, resulting in the formation of steps (Milne, 1982b; Grant et al., 1990). Thus, in contrast to the situation on bars, flow energy here in the channel is far higher and more concentrated. Step–pool formation could thus be a function of channel width; these features rarely extend into the wider, lower-energy bar zones.

It is questionable how far step–pool systems are dependent on channel pattern (Leopold et al., 1964, Keller and Melhorn, 1978) since the 150-m-long reach within the Lainbach test site is nearly straight and still has a well-developed step–pool system. Step–pools may be stronger indicators of sediment supply starvation than channel curvature in straight reaches. A large number of pools in the upper reaches of the Schmiedlaine are influenced by bedrock and the deficiency of sediment along this straight bedrock reach may also explains why step–pools are predominant there. Step–pool development may be only indirectly influenced by channel pattern as, for example, in the channel bends where there is a higher probability for coarse material deposition and its release into the main channel. Towards the lower depositional reaches, fewest step–pools occur, not only because of the oversupply of material but also because of decreasing gradients. Gradients determine the distance from step to step and step–pools form preferentially where gradients are steep.

Experiments under natural conditions (Ergenzinger and Stüve, 1989) indicate that sediment transport occurs from pool to pool in the Lainbach, suggesting that the pools are more active than the step regions during high flows and that the step regions are actually more stable (Ergenzinger and Schmidt, 1990). Observations and measurements during low flows have shown that shear stress distribution is usually similar in the step and pool areas since the steeper steps have shallower flows which compensate for the gentler pools with the higher depths. During an average flood, there is considerable difference in shear stress between step and pool areas. In contrast, high flood flows have a tendency to equalize the velocity in both areas, and this provides the conditions for step re-shaping.

The development of a step–pool system in the Lainbach and the Schmiedlaine was observed between 1990 and 1993. During a catastrophic flood/hyperconcentrated flow event with a 150 year recurrence interval (de Jong, 1992a), most of the Lainbach and the lower reaches of the Schmiedlaine, formerly step–pool, became braided depositional systems under heavy sediment input and general smoothing of the long profile. The long profile was straightened due to the erosion of some steps and infilling of all pools. Within just two weeks of the flood event, during a period of extreme re-adjustment, both the Lainbach and the Schmiedlaine started resembling a step–pool system again. This rapid re-establishment has been observed by other workers (Whittaker and Jaeggi, 1982; Sawada et al., 1983). Step–pool formation therefore occurs after the passage of high flows and coarsening of the bed. In the Schmiedlaine, nearly the same volumes of sediment deposited during the flood were transported away during that period. This was reconstructed from changes in the cross-profiles (Figure 4.2).

For a step–pool system to come into existence, erosion and sediment starvation are a prerequisite. In the Schmiedlaine and Lainbach, the starvation of sediment supply meant that the lower reaches developed higher shear stress due to an increase in gradient and it was possible for the finer material to be winnowed out. This eventually uncovered the coarse boulders, most of which had not travelled very far, to give rise to the original step–pool system. Grant et al. (1990) similarly found that their step–pool system formed in the absence of sediment transport. High sediment transport rates destabilized the sediment aggregations that formed candidates for step formation.

Bars

Although bar forms have been commonly described in sandy or gravelly meandering rivers, little attention has been given to the role that obstructions and bends play in controlling geomorphic forms in coarse-grained environments (Figure 4.13; Lisle, 1986). Bars are mainly deposited in the middle to lower reaches of the Schmiedlaine where the channel has high width to depth ratios (Chang, 1980). Deposition of point bars can be directly related to bend curvature reflecting the role that sharp bends play in arresting sediment under lower energy conditions. Other causes of bar deposition relate to obstructions caused by large, oversized boulders or bedrock promontories as in the case of the re-attachment bar, or simply to a decreased energy zone related to channel bifurcation. The role that wooden debris plays in obstructing sediment is treated separately under the section on log jams.

Measurements of grain size at selected locations showed that the sediment size of gravel was linked to the radius of curvature as in Milne (1982a). That is, the largest radius of curvature exhibited the finest grain size whereas the smallest radius of curvature was related to the coarsest sediment. In bends with the smallest radius of curvature a suitable environment was provided for the accumulation of logs behind large boulders, causing features such as log jams to develop. The formation of these alternating bars is strongly related to the radius of bend curvature not only due to the varying velocity and angle of attack (Lisle, 1986), but also due to the effects of superelevation differences of the right and left water levels. During the extreme flood of 30 June 1990 in the Lainbach Valley (de Jong, 1992a), the lateral difference in water level attained a gradient of up to 0.06, compared to the average of 0.04 in the long profile. These extreme gradients have very significant effects on flow energy and the capacity to transport sediment and explain the extent of these bar forms and their variation between the inner and outer bend (as in the case of the cobble berms) throughout the Schmiedlaine. In the bends, the bars are formed as a result of bend-induced secondary circulations (Nelson and Smith, 1989).

The changes in bedform types have to be related to the hydraulic geometry, i.e. the changing downstream relations between width and depth which essentially influence deposition (Schumm, 1960). Where there is a transition from low to high width–depth ratios, the channel shape changes accordingly from highly symmetrical to very asymmetrical. In the very wide meanders of the middle to lower reaches there is a tendency for a double symmetry to develop (de Jong, 1992a). In this case the geometry of the upper reach is bedrock-induced. Since it is mainly erosional, bed roughness and geometry will usually consist of single "grain" roughness or very large channel-filling boulders, as shown in cross-section L'M', Figure 4.2. In the lower reaches, form and unit roughness dominate; the channel bed consists of complex bedforms, mostly depositional in nature. In the upper reaches, erosion is far more effective since scour is possible to a greater depth for a small rise in stage. Due to the higher width to depth ratios in the lower reaches, velocity has far more impact on the channel bed, causing bedforms to develop accordingly.

In the Schmiedlaine, the complex hydraulic geometry of the cross-sections cause major differences in lateral water levels during large floods (see Takahashi, 1991, p. 94) and debris- and hyperconcentrated flows. Differences in the water levels need not even be restricted to sharp bends. In fact in the tributary Lainbach (see Figure 4.2) there were also measured water-level differences along the straight reach corresponding with secondary-flow cell development. In the bends the water-level difference is, of course, much larger so that the associated development of secondary circulation shapes laterally steeper sloping bars. Flow-cell development inhibits bar formation within their boundary of impact but, at the edges, bar forms will readily develop. Bars are thus associated with shallower flows where the diamond patterns of clusters suggest the predominance of shear waves.

The size, and especially the shape, of a large obstruction determines whether or not a pool or a bar will form downstream of it. Obstructions that did not project above the surface of maximum flow and that were deposited parallel to flow, not transverse to it, caused downstream bar development. Others that were outsized and formed a major "step", developed an erosive pool and bars upstream. In each case the bar length is dictated by the geometry of the bend; a bar only occurs within the low-energy zone before the next bend commences. This forms the basis of Lisle's (1986) model of the effects of bends and obstructions in stabilizing or arresting bar migration downstream. In the actively reworked bars of the Schmiedlaine, however, the configuration of the bar and channel does change and may cause the bars to occupy positions extending further up or downstream from their previous position.

In contrast to other work, we find that approach angles of flow are more influential than the actual obstruction in dictating bar development. This was observed on the test bar as it changed between 1990 and 1992. As the upstream thalweg changed, so did the position of the channel and bar, swinging away from its formally optimal position parallel to the bend. Thus the development of bedform roughness depends on the type of flow, sediment transport and the character of flow over the wavy surface (Allen, 1982a).

Log jams

The role of log jams as roughness elements is often neglected. Log jams are found exclusively in the sharpest channel bends of the Schmiedlaine (Figure 4.13) or downstream of a major debris flow input (Mosley, 1981). They form during extreme events and may also take on the shape of a debris end-lobe, a tongue-like accumulation of large particles. Log jams and end-lobes present the problem of differentiating between the competence of the flood itself (Komar, 1987) and the local slope or debris channel inputs. Since the material is so diverse in the Schmiedlaine, source determination is difficult.

Log jams are usually the result of a large log being grounded diagonally across the bend. Other logs and large boulders soon become trapped behind. The logs will arrange themselves parallel to the obstructing log. These depositional features were studied in detail for the Schmiedlaine. Figure 4.2 indicates how log jams run in tandem with major areas of deposition. These major bedforms are either dependent on the local sediment supply or on the storage of large material from extreme events (Grant *et al.*, 1990). Log jams have an important influence on channel morphology and are sediment storage sites (Keller and Tally, 1979). Since log jams cause large energy losses, they may actually stabilize the stream.

CONCLUSIONS

It is important to evaluate the function of micro- and macro-bedforms on the river bed in terms of river-bed stability. From these detailed studies of the spatial variability of roughness, it has become evident that the sorting of particles of similar shape and size into roughness assemblages such as clusters, varies according to the nature of the obstacle clast and the streamlines induced by cluster geometry. These factors are important in the stabilization of clusters. Cluster bedforms are particle-size and shape selective and this has important consequences for the regular patterns of roughness formation, repeated from year to year.

Other factors influencing the arrangement of roughness are the channel-bed gradient and curvature which ultimately control flow orientations. The analysis of different roughness scales indicates that the form sequences are not random (Grant *et al.*, 1990). Although there is considerable complexity in the different roughness systems of the Schmiedlaine, channel curvature clearly is essential in determining the length and type of bedform.

Since the grain-size distribution is highly variable downstream (Milne, 1982b) and is greatly influenced by channel curvature, the characteristics of roughness and geometry cannot simply be related to downstream fining. Log jams and debris flow end-lobes disturb downstream fining patterns.

The channel is stable only in those areas where it is unable to migrate due to bedrock restrictions. In all other cases, the channel dimensions and position vary considerably, particularly in the middle to lower, highly meandering reaches with high width to depth ratios. The thalweg attempted to straighten its route during the extreme flood but had a more meandering form in the preceding years; form and system roughness adjusted accordingly. Large roughness developed during the extreme events in positions not reworked during smaller floods. The different flood routing during the extreme flood can be explained in terms of sediment enrichment and energy losses during the extreme event, in contrast to the sediment-poor and energy-intensive processes during the smaller floods. The evidence gathered from channel roughness forms, both longitudinally and cross-sectionally, suggests that the micro and macro bedform developments depend, not only on channel curvature, but also on the combined effect of different types of discharge and sediment transporting events. Thus the amalgamation of perennial bedforms and high-energy features has to be clearly differentiated in river morphology studies.

Roughness development should not be considered only in one dimension. If bend curvature, flow orientations, flow depth, grain size and gradients are all to be included in roughness descriptions, then the river system has to be viewed in three dimensions. The interactions among these variables are not only important in determining the type of roughness, it is also the type of roughness that determines the geometry of the reach.

ACKNOWLEDGEMENTS

This study was part of the first author's PhD thesis while in receipt of a scholarship from the Studienstiftung des deutschen Volkes. The funding for fieldwork and equipment was made possible through the Deutsche Forschungs Gemeindschaft (German National Research Council) projects on "Bedload dynamics at Squaw Creek, Montana" and "The fluvial morphodynamics of the Lainbach in the Upper Quarternary".

REFERENCES

Allen, J. R. L. 1968. The nature and origin of bedform hierachies. *Sedimentology*, **10**, 161–182.
Allen, J. R. L. 1982a. Bedforms in supercritical and related flows: transverse ribs, rhomboidal features, and antidunes. In: Allen, J. R. L. (ed.), *Sedimentary Structures: Their Character and Physical Basis*, Elsevier, Amsterdam, pp. 383–416.
Allen, J. R. L. 1982b. Orientation of particles during sedimentation: shape fabrics. *Sedimentary Structures: Their Character and Physical Basis*, Elsevier, Amsterdam, pp. 179–235.
Andrews, E. D. 1979. Hydraulic adjustment of the East Fork River, Wyoming, to supply of sediment. In: Rhodes, D. D. and Williams, G. P. (eds), *Adjustments of the Fluvial System*, Kendall Hunt, Dubuque, Iowa, pp. 69–94.
Bagnold, R. A. 1956. The flow of cohesionless grains in fluids. *Transactions Royal Society London*, **249**, 235–297.
Becht, M. 1989. Der Einfluß von Muren, Schneeschmelze und Regenniederschlägen auf die Sedimentbilanz eines randalpinen Wildbachgebietes. *Die Erde*, **120**, 189–202.
Billi, P. 1988. A note on cluster bedform behaviour in a gravel-bed river. *Catena*, **15**(5), 473–481.
Boothroyd, J. C. and Ashley, G. 1975. Process, bar morphology and sedimentary structures on braided outwash fans, northeastern Gulf of Alaska. In: Jopling, A. V. and McDonald, B. C. (eds), *Glaciofluvial and Glaciolacustrine Environments*, Society of Economic Paleontologists and Mineralogists, Special Publication, pp. 193–222.
Brayshaw, A. C. 1984. The characteristics and origin of cluster bedforms in coarse-grained alluvial channels. In: Koster, C. H. and Steel, R. H. (eds), *Sedimentology of Gravels and Conglomerates*, Canadian Society of Petroleum Geologists, pp. 77–85.

Brayshaw, A. C. 1985. Bed microtopography and entrainment thresholds in gravel-bed rivers. *Geological Society of America Bulletin*, **96**, 218–223.
Brayshaw, A. C., Frostick, L. E. and Reid, I. 1983. The hydrodynamics of particle clusters and sediment entrainment in coarse alluvial channels. *Sedimentology*, **30**, 137–143.
Bridge, J. S. and Jarvis, J. 1982. The dynamics of a river bend: a study in flow and sedimentary processes. *Sedimentology*, **29**, 499–541.
Brush, L. M. 1965. Experimental work on primary sedimentary structures. In: Middleton, G. V. (ed.), *Primary Sedimentary Structures and their Hydrodynamic Interpretation*, Society of Economic Paleontologists and Mineralogists, Spec. Publ. 12, Tulsa, USA, pp. 17–25.
Buffington, J. M., Dietrich, W. E. and Kirchner, J. W. 1992. Friction angle measurements on a naturally formed gravel streambed: implications for critical boundary shear stress. *Water Resources Research*, **28**(2), 411–425.
Carling, P. A. 1989. Hydrodynamic models of boulder berm deposition. *Geomorptology*, **2**, 319–340.
Chang, H. H. 1980. Geometry of gravel streams. *Journal Hydraulics Division, ASCE*, **106**(HY9), 1443–1456.
Chin, A. 1989. Step pools in stream channels. *Progress in Physical Geography*, **13**, 391–407.
Dal Cin, R. 1968. Pebble clusters, their origin and utilization in the study of paleo-currents. *Sedimentary Geology*, **2**, 233–242.
de Jong, C. 1991. A re-appraisal of the significance of obstacle clasts in cluster bedform dispersal. *Earth Surface Processes and Landforms*, **16**(8), 737–744.
de Jong, C. 1992a. The dynamics of a catastrophic flood/multiple debris flow in the Schmiedlaine, S Germany. In: Walling, D. E., Davies, T. R. and Hasholt, B. (eds), *Erosion, Debris Flows and Environment in Mountain Regions*, IAHS, Chengdu, 209, pp. 237–245.
de Jong, C. 1992b. Measuring changes in micro and macro roughness on mobile gravel beds. In: Bogen, J., Walling, D. E. and Day, T. (eds), *Erosion and Sediment Transport Monitoring Programmes in River Basins*, IAHS, Oslo, 210, pp. 31–40.
de Jong, C. 1993. Temporal and spatial interactions between river bed roughness, geometry, bedload transport and flow hydraulics in mountain streams – examples from Squaw Creek, Montana, USA, and Lainbach/Schmiedlaine, Upper Bavaria, Germany. PhD thesis, Free University of Berlin.
Diepenbroek, M. 1992. Die Beschreibung der Korngestalt mit Hilfe der Fourier-Analyse. PhD thesis, Free University of Berlin.
Diepenbroek, M. and de Jong, C. 1993. Quantification of textural particle characteristics by image analysis – examples from active and paleo-surfaces in steep, coarse-grained mountain environments. In: Ergenzinger, P. and Schmidt, K. (eds), *Dynamics and Geomorphology of Mountain Rivers*, Springer, Benediktbeuern, pp. 301–314.
Diepenbroek, M., Bartholomä, A. and Ibbeken, H. 1992. How round is round? A new approach to the topic "roundness" by Fourier grain shape analysis. *Sedimentology*, **39**, 411–422.
Dobin, K. 1985. *Geologische Karte von Bayern 1:25 000*. Blatt Kochel. Bayeryisches Geologisches Landesamt, München.
Ergenzinger, P. 1992. River bed adjustments in a step–pool system: Lainbach, Upper Bavaria. In: Billi, P., Hey, R. D., Thorne, C. R. and Tacconi, P. (eds), *Dynamics of Gravel-Bed Rivers*, Wiley, Chichester, pp. 415–430.
Ergenzinger, P, and Jüpner, R. 1992. Using COSSY (CObble Satellite SYstem) for measuring the effects of lift and drag forces. In: Bogen, J., Walling, D. E. and Day, T. (eds), *Erosion and Sediment Transport Monitoring Programmes in River Basins*, IAHS, Oslo, 210, pp. 41–49.
Ergenzinger, P. and Schmidt, K. H. 1990. Stochastic elements of bedload transport in a step–pool mountain river. In: Sinniger, O. and Monbaron, M. (eds), *Hydrology in Mountain Regions. I: Artificial Reservoirs. Water and Slopes*, Special Publication, IAHS, Lausanne, 194, pp. 39–46.
Ergenzinger, P. and Stüve, P. 1989. Räumliche und zeitliche Variabilität der Fliesswiderstände in einem Wildbach: Der Lainbach bei Benediktbeuern in Oberbayern. *Göttinger Geographische Abhandlungen*, **86**, 61–79.

Felix, P. K., Priesmeier, H. V. and Wilhelm, F. 1988. Abfluß in Wildbächen. Untersuchungen im Einzugsgebiet des Lainbaches bei Benediktbeuem. *Münchner Geographische Abbhandlungen*, Band 6.

Grant, G. E., Swanson, F. J. and Wolman, M. G. 1990. Pattern and origin of stepped-bed morphology in high gradient streams, Western Cascades, Oregon. *Geological Society of America Bulletin*, **102**, 340–352.

Gustavson, T. C. 1974. Sedimentation on gravel outwash fans, Malaspina Foreland, Alaska. *Journal of Sedimentary Petrology*, **44**, 374–389.

Hassan, M. A. and Reid, I. 1990. The influence of microform bed roughness elements on flow and sediment transport in gravel bed rivers. *Earth Surface Processes and Landforms*, **15**(8), 739–750.

Hodges, W. K. 1982. Hydraulic characteristics of a badland pseudo-pediment slope system during simulated rainstorm experiments. In: Bryan, R. and Yair, A. (eds), *Badland Geomorphology*, Geo Books, Norwich, pp. 127–153.

Ibbeken, H. and Schleyer, R. 1986. Photo-sieving: a method for grain size analysis of coarse-grained, unconsolidated bedding surfaces. *Earth Surface Processes and Landforms*, **11**, 59–77.

Karcz, I. 1968. Fluviatile obstacle marks from the wadis of the Negev (Southern Israel). *Journal of Sedimentary Petrology*, **38**, 1000–1012.

Karcz, I. 1981. Reflections on the origin of small-scale longitudinal streambed scours. In: Morisawa, M. A. (ed.), *Fluvial Geomorphology*, George Allen and Unwin, London, pp. 149–173.

Keller, A. and Melhorn, N. 1978. Rythmic spacing and origin of pools and riffles. *Geological Society of America Bulletin*, **89**, 723–730.

Keller, E. A. and Tally, T. 1979. Effects of large organic debris on channel form and fluvial processes in the coastal Redwood environment. In: Rhodes, D. D. and Williams, G. P. (eds), *Adjustments of the Fluvial System*, Kendall Hunt, Dubuque, Iowa, USA, pp. 169–196.

Klingeman, P. C. and Matin, H. 1993. Incipient motion in gravel-bed rivers. In: Shen, W., Su, S. T. and Wen, F. (eds), *Hydraulic Engineering 1993, ASCE*, Vol. 1., San Francisco, pp. 707–712.

Komar, P. D. 1987. Selective grain entrainment by a current from a bed of mixed sizes: a reanalysis. *Journal of Sediment Petrology*, **57**, 203–211.

Koster, E. H. 1978. Transverse ribs: their characteristics, origin and paleohydraulic significance. In: Miall, A. D. (ed.), *Fluvial Sedimentology*, Canadian Society of Petroleum Geologists, pp. 161–186.

Laronne, J. B. and Carson, M. A. 1976. Interrelationships between bed morphology and bed-material transport for a small, gravel-bed channel. *Sedimentology*, **23**, 67–85.

Lawrence, G. A. 1987. Steady flow over an obstacle. *Journal of Hydraulic Engineering*, **113**(8), 981–991.

Leopold, L. B. 1992. Sediment size that determines channel morphology. In: Billi, P., Hey, R. D., Thorne, C. R. and Tacconi, P. (eds), *Dynamics of Gravel-Bed Rivers*, Wiley, Chichester, pp. 297–312.

Leopold, L. B. and Wolman, M. G. 1957. River channel patterns: braided, meandering and straight. *Professional Paper United States Geological Survey*, **282B**, 34–85.

Leopold, L. B., Wolman, M. G. and Miller, J. P. 1964. *Fluvial Processes in Geomorphology*, W. H. Freeman, San Francisco.

Lisle, T. 1986. Stabilization of a gravel channel by large streamwise obstructions and bedrock bends, Jacoby Creek, northwestern California. *Geological Society of America Bulletin*, **97**, 999–1011.

McDonald, B. C. and Banerjee, I. 1971. Sediments and bedforms on a braided outwash plain. *Canadian Journal of Earth Sciences*, **8**, 1282–1301.

Milne, J. A. 1982a. Bed-material size and the riffle-pool sequence. *Sedimentology*, **29**, 267–278.

Milne, J. A. 1982b. Bedforms and bend-arc spacing of some coarse-bedload channels in upland Britain. *Earth Surface Processes and Landforms*, **7**, 227–240.

Mosley, M. P. 1981. Semi-determinate hydraulic geometry of river channels, South Island, New Zealand. *Earth Surface Processes and Landforms*, **6**, 127–137.

Naden, P. M. and Brayshaw, A. C. 1987. Small and medium scale bedforms in gravel-bed rivers. In: Richards, K. S. (ed.), *River Channels: Environment and Process*, Basil Blackwell, Oxford, pp. 249–271.

Nelson, J. M. and Smith, J. D. 1989. Mechanics of flow over ripples and dunes. *Journal of Geophysical Research*, **24**, 8146–8162.

Osterkamp, W. R. 1978. Gradient, discharge and particle size relations of alluvial channels in Kansas with observations on braiding. *American Journal of Science*, **278**, 1253–1268.

Reid, I. and Frostick, L. E. 1986. Dynamics of bedload transport in Turkey Brook, a coarse-grained alluvial channel. *Earth Surface Processes and Landforms*, **11**, 143–155.

Reid, I., Frostick, L. E. and Brayshaw, A. C. 1992. Microform roughness elements and the selective entrainment and entrapment of particles in gravel-bed rivers. In: Billi, P., Hey, R. D., Thorne, C. R. and Tacconi, P. (eds), *Dynamics of Gravel-Bed Rivers*, Wiley, Chichester, pp. 253–276.

Richards, K. S. 1978. Channel geometry in the riffle pool system. *Geografiska Annaler*, **60A**, 345–354.

Robert, A. and Richards, K. 1988. On the modelling of sand bedforms using the semivariogram. *Earth and Planetary Science Letters*, **13**, 459–473.

Rubin, D. M., Schmidt, J. C. and Moore, J. M. 1990. Origin, structure and evolution of a re-attachment bar, Colorado River, Grand Canyon, Arizona. *Journal of Sedimentary Petrology*, **60**(6), 982–991.

Sawada, T., Ashida, K. and Takahashi, T. 1983. Relationship between channel pattern and sediment transport in a steep gravel bed river. *Zeitschrift für Geomorphologie*, N.F. Suppl. Band, **46**, 55–66.

Schumm, S. A. 1960. The shape of alluvial channels in relation to sediment type. *US Geological Survey Professional Paper*, **352**(B), 17–30.

Shen, H. W., Fehlman, H. M. and Mendoza, C. 1990. Bedform resistance in open-channel flow. *Journal of Hydraulic Engineering*, **116**(6), 799–815.

Sutherland, A. J. 1967. Proposed mechanism for sediment entrainment by turbulent flows. *Journal of Geophysical Research*, **72**, 6183–6194.

Sutherland, A. J. 1987. Static armour layers by selective erosion. In: Thorne, C. R., Bathurst, J. C. and Hey, R. D. (eds), *Sediment Transport in Gravel-bed Rivers*, Wiley, Chichester, 243–268.

Tait, S. J. and Willets, B. B. 1992. Characterisation of armoured bed surfaces. *Proceedings International Grain Sorting Seminar*, pp. 207–225.

Takahashi, T. 1991. Debris flow. In: *IAHR Monograph, Fluvial Hydraulics*, A. A. Balkema, Rotterdam, p. 165.

Teisseyre, J. G. 1977. Pebble clusters as a directional structure in fluvial gravels: modern and ancient examples. *Geologia Sudetica*, **12**, 79–92.

Whittaker, J. G. and Jaeggi, M. N. R. 1982. Origin of step–pool systems in mountain streams. *Journal of Hydraulics Division, ASCE*, **108**(HY6), 737–758.

Wiberg, P. L. and Smith, J. D. 1987. Calculations of critical shear stress for motion of uniform and heterogeneous sediments. *Water Resources Research*, **23**(8), 1471–1480.

Wolman, M. G. 1954. A method for sampling coarse river-bed material. *Transactions American Geophysical Union*, **35**(6–1), 951–956.

Zgheib, P. W. 1990. Large bed element channels in steep mountain streams. In: Sinniger, O. and Monbaron, M. (eds), *Hydrology in Mountain Regions II: Artificial Reservoirs. Water and Slopes*, Special Publication, IAHS, Lausanne, 194, pp. 277–283.

5

Effective Discharge for Bedload Transport in a Subhumid Mediterranean Sandy Gravel-Bed River (Arbúcies, North-East Spain)

R. J. BATALLA AND M. SALA

Departament de Geografia Física, Universitat de Barcelona, Spain

ABSTRACT

The main objective of this paper is to examine the relation between the geomorphic dominant discharge (bankfull) and the effective discharge for bedload transport in a sandy gravel-bed river in a humid Mediterranean forested granitic drainage basin (Arbúcies, north-east Spain). The drainage area is 114 km^2. Discrete bedload samples and flow data were obtained weekly as well as during flood events during 1991 and 1992. Bankfull stage was estimated by means of field mapping and physical evidence. Bedload transport occurs as almost constant movement of the sandy bed material in the channel, even during low discharges. The annual average bedload yield is 46% of the mean annual total yield including solutes, and represents more than 65% of the annual solid yield. Calculations show that the bedload yield is dominated by events of high frequency and moderate magnitude associated with the bankfull discharge. These events occur 2.2% of the time and carry 31% of the annual bedload yield. Effective discharge for bedload transport in this river is, thus, a relatively frequent event of identical magnitude to the bankfull (geomorphic dominant) discharge.

INTRODUCTION

Information about bedload transport is very useful to evaluate erosion rates, to describe sediment dynamics, and to assess the geomorphic implications of sediment transport. This information can also be very valuable for civil engineering purposes. However, bedload measurements in relation to sediment yield are not often available. In sand-bed channels, bedload may constitute a small proportion (1–20%) of the total load (Lane and Borland, 1951; Simons and Senturk, 1977). Similar values are reported from gravel-bed rivers (McPherson, 1971; Dietrich and Dunne, 1978).

Bedload movement is both episodic and discontinuous. The significance of a given discharge capable of transporting bedload over a period of years is determined by two main factors: event magnitude and frequency. The concepts of magnitude and frequency of geomorphic processes were developed by Wolman and Miller (1960). One of the principles they presented is that, in rivers, the effective geomorphic force is a relatively frequent event. They suggested that in humid temperate environments, the effective discharge for sediment transport would be very similar in magnitude to the dominant discharge controlling channel morphology. However, Benson and Thomas (1966) concluded that, for most rivers, the effective discharge is much less than bankfull discharge and lies somewhere between the mean annual flow and the discharge indicated by Wolman's investigations.

Few attempts have been made to relate bedload transport to the geomorphic dominant discharge in order to establish the relation between bankfull and transport-effective discharges. Pickup and Warner (1976) found that the effective discharge for bedload transport has a shorter return period than the bankfull stage for small gravel-bed, clay-bank streams near Sydney, Australia. Further, Andrews (1980) has shown for the total load that the effective discharge is a relatively frequent event exceeded between 1 and 10 days per year (0.35–3% of the time) in the Yampa River basin, USA. Ashmore and Day (1988) indicated that the duration of the effective discharge is related to drainage area, as it reflects differences in the form of the flow-duration curves (Pickup and Warner, 1976).

A sediment transport investigation has been undertaken since 1991 in the basin of the Arbúcies River, Catalan Coastal Ranges, north-east Spain. The work has concentrated on the transport of the sand bedload. The main purpose of this paper is to examine the relation between the geomorphic dominant discharge (bankfull discharge) and bedload transport in a Mediterranean sandy gravel-bed river, assessing its magnitude and frequency and its contribution to the sediment yield. Relations between discharge, bedload transport rates and the size of the mobile sediment are also examined.

THE BASIN

The Arbúcies River is one of the main tributaries of the Tordera River. The 114 km^2 basin is located in the Catalan Coastal Ranges, in the north-eastern part of the Iberian Peninsula (Figure 5.1). Average slope is 5% along the basin while the slope at the study reach in the lower part of the basin is 0.95%. Granitic bedrock covers more than 90% of the basin area, with granodiorite the most common type. There is a great depth of free-draining weathered granitoids at all sites indicated by resistivity soundings (Cervera, 1986) consisting of up to 22 m of boulders, and especially sand and fine material. There are also small areas of schist, micaschist and gneiss. Quaternary deposits consist of three Holocene terraces where sand and fine gravels are predominant. Mean annual rainfall is 984 mm and mean annual evaporation is 637 mm. Vegetation mainly consists of an evergreen-oak woodland that covers 97% of the basin. The analysis of discharge for the period 1967–1992 (Batalla, 1993) shows that there is continuous streamflow except for 2% of the time, which gives an average of six to seven drought days per year. The mean flow is 1.1 m^3 s^{-1}. Mean

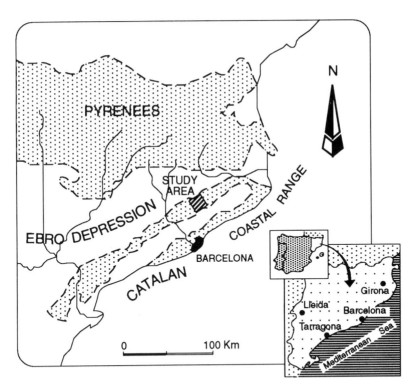

FIGURE 5.1   Location of the Arbúcies drainage basin, Catalan Coastal Ranges, north-east Spain

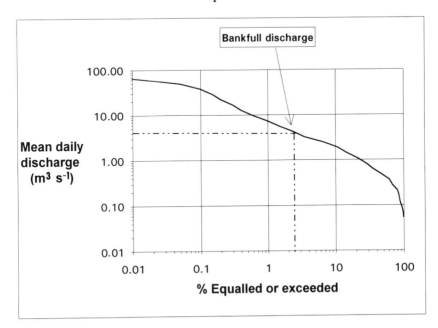

FIGURE 5.2   Flow-duration curve of the Arbúcies River, gauging station 56, based on daily data for the period 1967–1992 (Batalla, 1993). Mid-point discharge of bankfull stage is indicated

base-flow is 0.5 m³ s⁻¹. The Lane and Lei (1950) streamflow variability index of the flow-duration curve (Figure 5.2) is 0.39. Floods can reach 65 m³ s⁻¹ (recurrence interval of 50 years). The flash flood magnitude index (Beard, 1975; Baker, 1977) for the Arbúcies River is 0.52. River-bed material is mainly poorly sorted sandy gravel with D_{50} and D_{95} equal to 2.2 mm and 71 mm respectively. Textural changes are also observed between the river-bed material and the sediment of the banks where D_{50} is 0.7 mm and D_{95} is 6 mm.

DATA COLLECTION

Bedload and flow data were obtained weekly as well as during flood events during 1991–1992. Bedload material was sampled using a pressure-difference Helley-Smith sampler with a 76.2 mm intake, 0.0059 m² orifice, 0.45 mm mesh, and 15 kg bag capacity. For particle sizes larger than 0.50 mm and smaller than 16 mm, sediment trapping efficiency of the sampler may reach 100%, irrespective of the changes in transport rates. Efficiency for particles larger than 16 mm drops to less than 70% (Emmett, 1979). Measurements of bedload transport in the Arbúcies River were not corrected since the proportion of particles larger than 16 mm were very often negligible (less than 1%) and, thus, the final computation of bedload discharge was not affected by the Helley-Smith trapping efficiency. At a site integrated bedload samples were collected every metre across the channel section, 50 m upstream from gauging station number 56. Sampling time ranged between 5 and 10 minutes.

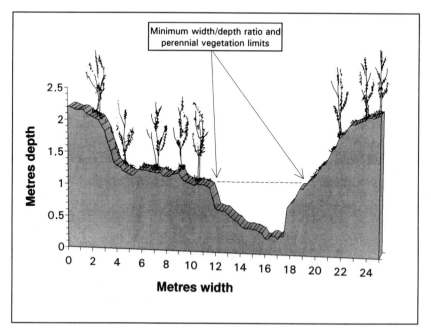

FIGURE 5.3 Cross-section upstream of gauging station number 56, Arbúcies River, indicating field evidence of bankfull stage, such as: vegetation lower limit and minimum width–depth ratio (trees are not vertically scaled)

FREQUENCY AND CHANNEL ADJUSTMENTS TO THE BANKFULL DISCHARGE

Bankfull stage (Figure 5.3) was estimated by means of field mapping and physical evidence, such as the lower vegetation limit (Schumm, 1960), and the minimum width/depth ratio (Wolman, 1955; 9.3 for the Arbúcies River). The flow-duration curve (Figure 5.2) shows that bankfull discharge (4.0 m^3 s^{-1}) is equalled or exceeded 2.2% of the time, or about seven days per year. This result is based on daily data for the period 1967–1992 obtained at the gauging station number 56 (Batalla, 1993). The maximum width and depth at bankfull stage are 7.5 m and 0.8 m respectively. Mean depth of the cross section is 0.51 m and mean velocity is 1.1 m s^{-1}.

Preliminary results obtained from magnetized scour-chains installed during 1992 upstream of gauging station 56 indicate that, during flood events, processes of scour and fill are highly active in this sandy gravel-bed river. At bankfull stage at least 50 cm of bed material are scoured and deposited during such events. As Pickup and Warner (1976) stated, short-term changes in channel characteristics may occur during the passage of a single flood. Moreover, since the Arbúcies channel in the main valley is in quasi-equilibrium, intermediate changes represent fluctuations about the average conditions.

RESULTS

Bedload transport

Over the study period, 72 samples were collected covering a wide range of discharges and sediment transport rates. The bedload mean transport rate (i_b) is 37.5 g m^{-1} s^{-1} (submerged weight), with a coefficient of variation of 178.3%. Bedload transport rates vary from 1.1 g m^{-1} s^{-1} under low discharges to 279.1 g m^{-1} s^{-1} during bankfull discharges. Figure 5.4 shows that the data are very scattered with a general trend of increasing transport rates with discharge. The mean transport rate for discharges less than bankfull level is 28.2 g m^{-1} s^{-1} with a coefficient of variation of 102.3%. On the other hand, the mean transport rate at bankfull stage is 124.5 g m^{-1} s^{-1} with a coefficient of variation of 82.3%. Measured changes in bedload transport rates in the Arbúcies River at bankfull stage vary between 80% and 600% under the same flow conditions. Because of technical difficulties, no representative data above bankfull stage were collected.

Figure 5.5 illustrates that there is only a weak relation between the median particle size of the mobile sediment and discharge, perhaps indicating that the fine sediment is in movement at almost any flow. Similar results were obtained from the maximum mobile sizes. However, a significant shift in the maximum mobile sizes is associated with the bankfull discharge. At that discharge the maximum size of the transported material increases almost one order of magnitude. This may indicate the destruction of the surface armoured layer and the release of large amounts of sediment ready to be transported. However, the role of an armour layer needs further investigation.

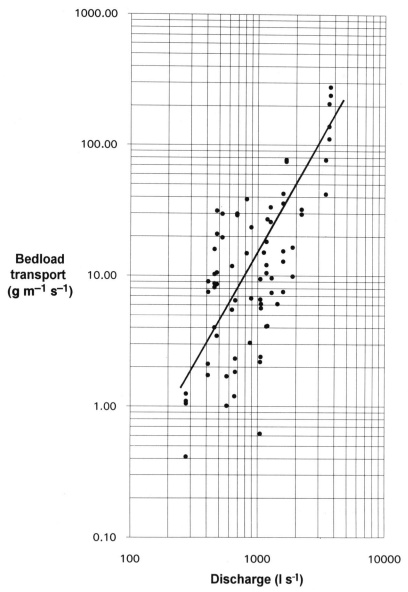

FIGURE 5.4 Relation between bedload transport rate and discharge in the Arbúcies River (equation 5.2)

Bedload yield

Bedload yield in the Arbúcies River is determined by the continuous movement of the sand fraction on the channel bed. Bedload transport was observed even during low discharges. Field observations showed that only 8 cm of water depth is needed to initiate the movement of fine sediment, a depth which is equalled or exceeded 80% of the time. Shear stress associated with 8 cm water depth is four times more

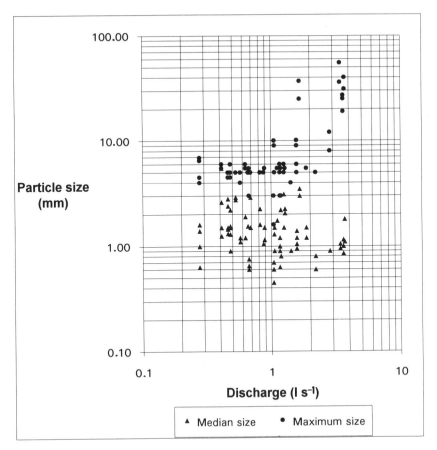

FIGURE 5.5 Relation between median and maximum particle size of the samples and discharge in the Arbúcies River

than that calculated using the entrainment function of Shields. This difference between the mean bed-shear stress and the critical shear stress can be attributed to the presence of bedforms at the initiation of motion.

The relation between the sediment transport rate (i_b) and unit discharge (q) was examined. This relation is statistically significant ($p < 0.01$) and given in equation (5.1):

$$i_b = 0.00071 \ q^{1.91} \qquad (r^2 = 0.47, \ n = 72) \qquad (5.1)$$

where q is l s^{-1} per unit width and i_b is g m^{-1} s^{-1} of submerged weight. Similar relations are reported from other studies (e.g. Henderson, 1966). To calculate the bedload yield of the basin, the relation between bedload transport rate and discharge (Q) was used:

$$i_b = 0.0008 Q^{1.47} \qquad (r^2 = 0.47, \ n = 72) \qquad (5.2)$$

The low correlation between the transport rate and discharge is reflected in the scattered nature of the data (Figure 5.4). The scatter in the data might be related to spatial and temporal variations of sand transport (Gomez et al., 1989), to changes in the hydraulic conditions, to the development and migration of bedforms, or to destruction and reformation of armour layers, especially during flood events.

Annual bedload yield was calculated using equation (5.2) and the flow duration curve (Figure 5.2). Calculations were based on Piest's (1964) and Walling's (1977, 1984) method. Statistical bias of the log-transformed least-squares relation between bedload transport rates and discharge has been adjusted by means of the correction factor of Ferguson (1986). The statistical deviation between measured and estimated transport rates has been calculated to be 1.65. Annual bedload yield in dry weight is 6665 tonnes (Table 5.1). The mean annual bedload yield of the river is more than 65% of the mean annual solid yield and 46% of the mean annual total yield, including solutes (Batalla, 1993). These values give a mean annual yield of 620 kg ha^{-1}.

The effective discharge was determined by dividing the flow into almost 20 classes, finding the duration of flow within each class, calculating the mean bedload discharge within the class using the significant statistical relation between discharge and bedload transport rate, and multiplying it by duration. Calculations show that bedload yield is dominated by events of high frequency and moderate magnitude

TABLE 5.1 Calculation of the mean annual yield of bedload material in the Arbúcies River, from the flow-duration curve (1967–1992) and the bias-corrected statistical relation between bedload transport rates and discharge, based on Piest's (1964) and Walling's (1977, 1984) method

Interval (%)	Mid-point	Duration (%)	Discharge (l s^{-1})	Transp. rate (g m^{-1} s^{-1})	Width (m)	Load* (tonnes)
(0–0.5)			No data above 4000 l s^{-1}			
(0.5–5)	2.2	2.2	3900	264.2	6.5	1209.1
(5–10)	7.5	5	2050	102.0	6.0	979.2
(10–15)	12.5	5	1500	64.2	5.0	513.9
(15–20)	17.5	5	1200	46.2	4.0	295.5
(20–25)	22.5	5	950	32.7	4.5	235.3
(25–30)	27.5	5	850	27.7	4.0	177.4
(30–35)	32.5	5	750	23.0	4.0	147.4
(35–40)	37.5	5	675	19.7	4.0	126.1
(40–45)	42.5	5	600	16.6	4.0	105.9
(45–50)	47.5	5	525	13.6	4.0	86.9
(50–55)	52.5	5	475	11.7	4.0	75.0
(55–60)	57.5	5	400	9.1	4.0	58.1
(60–65)	62.5	5	380	8.4	4.0	53.9
(65–70)	67.5	5	350	7.5	4.0	47.7
(70–75)	72.5	5	275	5.2	4.0	33.4
(75–80)	77.5	5	240	4.3	3.5	23.9
(80–85)	82.5	5	210	3.5	3.0	0.0
(85–90)	87.5	5	180	2.8	3.0	0.0
(90–95)	92.5	5	120	1.5	3.0	0.0
(95–100)	97.5	5	90	1.0	3.0	0.0

Mean annual yield (submerged weight) = 4168 tonnes
Mean annual yield (dry weight) = 6665 tonnes
(* Sediment yield = flow frequency · transport rate · bed width · 0.32)

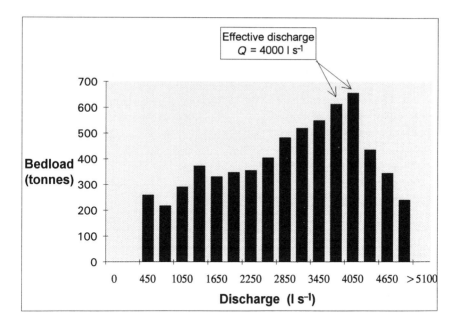

FIGURE 5.6 Histogram of mean annual bedload discharges, Arbúcies River

(Figure 5.6). The flow-duration curve shows that discharges associated to that of bankfull stage (around 4.0 m^3 s^{-1}) occur 2.2% of the time but yield about 31% of the total bedload material. Effective discharge for bedload transport is, therefore, close to bankfull stage.

DISCUSSION AND CONCLUSIONS

Arbúcies data are very scattered, possibly due to the variations of transport rates associated with migration of bedforms during sampling, especially at bankfull discharges. Presumably, sampling time was, on occasions, less than that required to identify the movement of the bedforms in the channel. This fact could partly affect the final accuracy of our results, especially those obtained during high flows.

The importance of the bedload contribution to the annual sediment yield in the study river is higher than those reported in other studies for sand-bed channels (Lane and Borland, 1951; Simons and Senturk, 1977) and for gravel-bed rivers (McPherson, 1971; Dietrich and Dunne, 1978). The values obtained in this study are similar to those obtained in coarse gravel-bed mountain rivers by Moosbruger (1957), Hayward (1980), Lauffer and Sommer (1982), Ergenzinger and Custer (1983), and Kjeldsen (1983). The relatively high yield of bedload in the Arbúcies basin is mainly due to the high frequency of discharges that are capable of moving large amounts of sand.

The relation between the effective discharge for bedload transport and the dominant discharge for channel morphology appears to be clear in the Arbúcies River. Arbúcies data show that the effective discharge for bedload transport is a relatively frequent event of moderate magnitude (exceeded seven days per year) close to that of the bankfull stage, as was suggested by Wolman and Miller (1960) and confirmed by Andrews (1980). According to Andrews (1980) the effective discharge for sediment transport is identical in magnitude to the bankfull discharge. However, Benson and Thomas (1966), Pickup and Warner (1976) and Nolan *et al.* (1987) show that the effective discharge is clearly below bankfull. Ashmore and Day (1988) claimed that the effective discharge depends on the drainage area. For small drainage basins of the Saskatchewan River, they found that extreme events are the most effective discharges, while frequent events are the most effective in terms of sediment transport for larger drainage areas. Thus, these authors presumed that the relation between effective and bankfull discharge is highly variable making complicated any generalizations on this theme.

The results obtained in the humid Mediterranean forested drainage basin of Arbúcies support the conclusion of Ashmore and Day (1988). As it is an individual research catchment, Arbúcies data do not reconcile the apparent discrepancies in terms of drainage area and effective discharge derived from their work. Nevertheless, Arbúcies data validate the effective discharge concept for fluvial sediment transport (bedload transport here) as it was essentially developed by Wolman and Miller (1960). Furthermore, these data specify the magnitude and duration of the effective discharge in humid Mediterranean environments characterized by interflow and base flow hydrologic processes, with low/moderate streamflow variability and flash flood magnitude indices, and where the drainage network may not be exclusively adjusted to extreme hydrological events.

ACKNOWLEDGEMENTS

This research was supported by grants from the Spanish Ministry of Education and Science and CIRIT (Generalitat de Catalunya). The hydrological data were supplied by the Junta d'Aigües, Generalitat de Catalunya. Alan Werritty of the University of St Andrews, UK, advised in all stages of this study. This paper benefits from discussions with Marwan Hassan of the University of British Columbia. We gratefully acknowledge the assistance of Dan Moore and Wayne Erskine whose suggestions have greatly improved the first version of this manuscript. We also wish to thank Peter Ergenzinger of the Freie Universität Berlin for his technical support. Darren Ham of the University of British Columbia reviewed the English version.

REFERENCES

Andrews, E. D. 1980. Effective and bankfull discharges of streams in the Yampa river basin, Colorado and Wyoming. *Journal of Hydrology*, **46**, 311–330.
Ashmore, P. E. and Day, T. J. 1988. Effective discharge for suspended sediment transport in streams of the Saskatchewan river basin. *Water Resources Research*, **24**(6), 864–870.
Baker, V. R. 1977. Stream-channel response to floods, with examples from Central Texas. *Geological Society of America Bulletin*, **88**, 1057–1071.
Batalla, R. 1993. *Contribucio del transport de sorres en el balanç de sediment d'una conca granitica Mediterrània*. Unpublished PhD thesis, Universitat de Barcelona, 199 pp.
Beard, L. R. 1975. *Generalized evaluation of flash-flood potential*. Texas University Center Research Water Resources Tech. Rept. CRWR-124, 27 pp.

Benson, M. A. and Thomas, D. M. 1966. A definition of dominant discharge. *Bulletin International Association Scientific Hydrology*, **11**, 76–80.

Cervera, M. 1986. Spatial variation of surface wash and erosion in the slopes of Santa Fe. *COMTAG Symposium, Excursion Guide*, 69–73.

Dietrich, W. and Dunne, Th. 1978. Sediment budget for a small catchment in mountainous terrain. *Zeitschrift für Geomorphologie* N.F. Suppl. Bd, **29**, 191–206.

Emmett, W. W. 1979. A field calibration of the sediment trapping characteristics of the Helley-Smith bedload sampler. *US Geological Survey Professional Paper*, **1139**.

Ergenzinger, P. and Custer, S. 1983. First experiences measuring coarse material bedload transport with a magnetic device. In: Sumer, B. M. and Muller, A. (eds), *Mechanics of Sediment Transport*. A. A. Balkema, Rotterdam, 223–227.

Ferguson, R. I. 1986. River loads underestimated by rating curves. *Water Resources Research*, **22**(1), 74–76.

Gomez, B., Naff, R. L. and Hubbell, D. W. 1989. Temporal variations in bedload transport rates associated with the migration of bedforms. *Earth Surface Processes and Landforms*, **14**, 135–156.

Hayward, J. A. 1980. Hydrology and stream sediments in a mountain catchment. *Spec. Publ. 17*, Tussock Grasslands and Mountain Lands Institute, Canterbury, New Zealand, 236 pp.

Henderson, F. M. 1966. *Open Channel Flow*, Macmillan, New York, 279 pp.

Kjeldsen, O. 1983. *Materialtransportundersokelser i Norske breelver 1981. Rep. 1–83*. Vassdragsdirektoratet Hydrologisk Avdeling, Oslo, 39 pp.

Lane, E. W. and Borland, W. M. 1951. Estimating bedload. *Transactions of the American Geophysical Union*, **32**(1), 121–123.

Lane, E. W. and Lei, K. 1950. Streamflow variability. *Transactions of the American Society of Civil Engineers*, **115**, 1084–1134.

Lauffer, H. and Sommer, N. 1982. Studies on sediment transport in mountain streams of the eastern Alps. *Proceedings 14th Congress International Commission on Large Dams*, Rio de Janeiro, Brazil, 431–453.

McPherson, H. J. 1971. Dissolved, suspended and bedload movement patterns in Two O'clock Creek, Rocky Mountains, Canada, Summer, 1969. *Journal of Hydrology*, **12**, 221–233.

Moosbruger, H. 1957. Le charriage et les debit solide en suspension des cours d'eau de montagnes. *Proceedings IAHS Toronto General Assembly*, IASH Publ. 38, pp. 243–259.

Nolan, K. M., Lisle, T. E. and Kelsey, H. M. 1987. Bankfull discharge and sediment transport in northwestern California, Erosion and sediment in the Pacific Rim, *IAHS Publ.*, **165**, 439–449.

Pickup, G. and Warner, R. F. 1976. Effects of hydrologic regime on magnitude and frequency of dominant discharge. *Journal of Hydrology*, **29**, 51–75.

Piest, R. F. 1964. Long-term sediment yields from small watershed. In: Land Erosion, Precipitation, Hydrometry, Soil Moisture. *Proceedings Berkeley General Assembly of IUGG, IAHS Publ.* **65**, 121–140.

Schumm, S. A. 1960. The shape of the alluvial channels in relation to sediment type. *US Geological Survey Professional Paper*, **352-B**.

Simons, D. B. and Senturk, F. 1977. *Sediment Transport Technology*, Water Resources Publications, Fort Collins, Colorado, 807 pp.

Walling, D. E. 1977. Limitations of the rating curve technique for estimating suspended sediment loads, with particular reference to British rivers. *IAHS Publ.* **22**, 34–48.

Walling, D. E. 1984. Dissolved loads and their measurements. In: Hadley, R. F. and Walling, D. E. (eds), *Erosion and Sediment Yield: Some Methods of Measurement and Modelling*. Geo Books, London, 111–177.

Wolman, M. G. 1955. The natural channel of Brandywine Creek, Pennsylvania. *US Geological Survey Professional Paper*, **271**.

Wolman, M. G. and Miller, J. P. 1960. Magnitude and frequency of forces in geomorphic processes. *Journal of Geology*, **68**, 54–74.

6

Within-Reach Spatial Patterns of Process and Channel Adjustment

S. N. LANE, K. S. RICHARDS

Department of Geography, University of Cambridge, UK

AND

J. H. CHANDLER*

Department of Civil Engineering, City University, London, UK

ABSTRACT

The average dimensions of a river reach in equilibrium may be determined by the hydrological regime of the basin and by the boundary conditions associated with the environment through which it flows. However, within the same reach, interaction amongst the morphology, sedimentology, flow hydraulics and sediment transport processes is manifest as a spatially distributed feedback, with the result that the river channel itself is in a state of continual change. Analysis of this process is demanding of data, often because very dynamic channels are required in order to collapse the timescale of adjustment. These channels are commonly morphologically complex (being, for example, multi-thread channels). Distributed information is required from within the study reach on the river-bed topography, sedimentology, flow parameters and bed-load transport. The required process data are difficult to obtain; and collecting topographic data at a comparable spatial and temporal resolution is particularly problematic.

Research on the proglacial stream of the Haut Glacier d'Arolla (south-west Switzerland) shows how such intensive field-data acquisition, coupled with numerical models of the adjustment of the incoming flow to the bed topography and sedimentology, can aid understanding of these relationships. Appropriate flow models are available for this purpose, although their application to natural channels at this scale of investigation is unusual. They vary in sophistication both in terms of the way in which they represent turbulence and the number of dimensions that are considered; in this situation a model with a relatively sophisticated turbulence closure and a two-dimensional solution is used. Oblique terrestrial analytical photogrammetry and rapid stream-bed survey are combined to construct a digital terrain model of the bottom elevation and water surface elevation of a braided section. With additional field data on bed roughness and flow discharge, this provides the necessary boundary conditions for application of the flow model.

---

*Current address: Department of Civil Engineering, Loughborough University of Technology, UK

River Geomorphology. Edited by Edward J. Hickin
© 1995 John Wiley & Sons Ltd

The comparison of distributed velocity measurements with model predictions allows some confidence to be placed in model results although a number of areas need further consideration. In particular, close attention must be paid to the problem of determining channel boundary roughness, as this has critical effects on the effectiveness of flow predictions.

INTRODUCTION

Previous studies that have endeavoured to link process measurements to river channel form have concentrated on ideas concerning the adjustment of landforms to the dominant process regime. Pickup (1976), for instance, describes the adjustment of alluvial channels in response to changing hydrological regime. However, recent intensive process-based studies of alluvial channels (e.g. Dietrich and Smith, 1983, 1984; Ashworth and Ferguson, 1986) have emphasized an alternative and smaller time and space scale for understanding alluvial channel dynamics. Emphasis on channel equilibrium is replaced by a consideration of channels in a state of continual change, as process events of varying magnitude continually act on an ever-changing morphology, whose shape itself influences the pattern of the formative processes. For instance, Naden and Brayshaw (1987) discuss both bedform construction by bedload transport, and bedform influences on bedload transport. The nature of these relationships can be summarized by the systems diagram in Figure 6.1 (Ashworth and Ferguson, 1986; Richards, 1988). Its essence is the feedback that exists in process-response systems where form at time $t=1$ influences process which produces a new form at time $t=2$.

Intensive process-based studies of this continual change have made use of particularly dynamic alluvial channels, such as those in glacierized catchments, where channel change is so rapid it permits direct field study. Such studies are particularly demanding of data, in terms of rapidity of collection, volume, and instrument sophistication necessary to provide the required process information. However, they have substantially increased understanding of processes *within* river channels (e.g. Ferguson *et al.*, 1989; Ashmore *et al.*, 1992). Nevertheless, despite the widespread recognition of the importance of the system described in Figure 6.1, process-based studies have yet to model in a quantitative manner the nature of the interaction between system components.

The first stage in quantitative modelling must consider the adjustment of the incoming flow to the topographical and sedimentological boundary condition. Appropriate models of turbulent fluid flow are available having been developed over the last 30 years by hydraulic engineers, although they have yet to be applied and tested to any significant degree in irregular natural river channels (ASCE, 1988). They require distributed information on the river-bed topography, bed sedimentology and discharge, and have the capability of predicting the spatial patterns of flow velocity and bed shear stress.

The purpose of this paper is to illustrate for a dynamic gravel-bed river how it is possible to combine intensive studies of within-channel form and process with computational models of turbulent fluid flow. After an introduction to the principles of modelling turbulent flow and a consideration of the flow model applied to this particular problem, a model application is described. Model predictions are

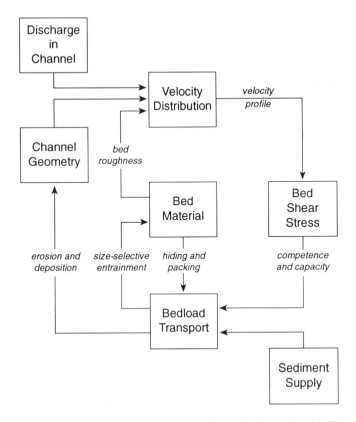

FIGURE 6.1 Form–process interrelationships in a dynamic river channel (from Ashworth and Ferguson, 1986, with a modification by Richards, 1988)

compared with distributed velocity observations. Once confidence in model predictions has been assured, the geomorphological implications of the results obtained are considered, and some suggestions are made as to how this research might be developed further.

MODELLING NATURAL RIVER FLOW

It is possible to show that turbulent open-channel flow satisfies at every point in space the Navier–Stokes equations for an incompressible fluid that are derived directly from the Principles of Conservation of Mass and Momentum. These are based upon the Eulerian transformation of Newton's Laws of Motion. Modifications are made necessary because the latter are formulated for fixed and rigid bodies, while a fluid represents a continually deforming continuum. In theory, it only remains to specify appropriate boundary conditions and to solve these equations to model the flow in a turbulent river channel. However, direct application of these equations results in a computational problem. The solution of this set of differential equations requires a three-dimensional computational grid that is large enough to cover the area of interest but which has spacings smaller than the smallest turbulent motion. The computation must be unsteady, utilizing a time-step smaller than that associated with the fastest eddies. It is possible to show that the number of grid nodes required shows a Reynolds number dependence (Emmons, 1970); with a Reynolds number of 10^5, not untypical in natural channel flows, the grid would require approximately 10^{21} nodes to cover a square metre. This is not feasible even with recent improvements in computer processing power. To overcome the problem, hydraulic engineers either consider a statistical theory of turbulence based upon empirical correlation among observations of the structure of turbulent river flow, or undertake a semi-empirical analysis to estimate the effects of turbulence on the mean flow properties.

Modelling natural river flow makes use of this latter method. There is some justification for it (Rodi, 1980) as it is mainly the large-scale turbulent motion that is responsible for the transport of momentum and it is therefore this that must be modelled in terms of its effects on the mean flow properties. This makes use of a technique called Reynolds-averaging which decomposes velocities and pressures in the Navier–Stokes equations into average values plus fluctuating components (Reynolds, 1895). No change occurs in the structure of the continuity equation, but averaging of the momentum equations introduces a new set of terms, the Reynolds shear stress tensor, which represent the transport of momentum that can be attributed to turbulence. These additional terms can never be zero in a turbulent flow and, as Rodi (1980) notes, in turbulent flows they tend to be much larger than their molecular counterparts. No additional equations have been developed to solve for these and the Reynolds-averaged equations do not close (Olson and Wright, 1990); the only alternative is to model their consequences using flow parameters that are either known or knowable using a variety of semi-empirical methods.

The turbulence problem

The simplest closure schemes are based upon a consideration of the statistical properties of turbulence and place emphasis upon empirical observations (Deissler, 1977). Although simple, they do not cope effectively with heterogeneous, well-developed turbulence in which initial conditions have no real effect on the final flow (Rodi, 1980). More frequently adopted is the Boussinesq (1877) concept, based upon a mixture of physical reasoning and dimensional analysis (Deissler, 1977). An analogy is made with molecular motion which leads to Stoke's velocity law in laminar flow (Rodi, 1980). Boussinesq proposed that the Reynolds shear stresses may be modelled as being proportional to the mean rates of strain, the proportionality coefficient being the eddy viscosity coefficient. The latter is not constant, but is a property of the flow (Olson and Wright, 1990) which can be determined to a good approximation in many situations (Rodi, 1980). A number of methods have been used to determine the eddy viscosity. The simplest are referred to as zero-equation models. These either prescribe the eddy viscosity (this is not really a turbulence model as it does not take into account changes in the local turbulence structure and hence it cannot describe correctly the details of the mean flow field) or make use of a mixing-length assumption (Prandtl, 1925), where it is proposed that the eddy viscosity is proportional to a mean fluctuating velocity and a mixing-length. The latter has seen widespread application to problems of river channel hydraulics, being the basis of the "law-of-the-wall" method used for calculating bed shear stress from velocity profiles (Carson, 1971), and being used in those attempts that have been made to model natural river channel flow (e.g. Whiting and Dietrich, 1991).

The main problem with zero-equation models is that they implicitly assume that turbulence is dissipated where it is generated, and in doing so fail to account for the transport of turbulence by the mean flow. This is a particular problem in those situations where the state of turbulence at a point is significantly influenced by turbulence generation elsewhere in the flow (e.g. locations downstream of a vortex shedding boulder). One-equation and two-equation models endeavour to take into account these "history" effects by modelling the velocity scale or the velocity and length scales respectively, in terms of the mean flow properties. The most widely adopted model in of hydraulic engineering is the two-equation k-ε model, based on this method. The velocity scale is related to the square root of the kinetic energy of the turbulent motion per unit mass (k). The rate of change of k is balanced by convective transport (due to the mean motion), diffusive transport (due to velocity and pressure fluctuations), the production of k (due to interaction of Reynolds stresses and mean velocity gradients) and the dissipation of k (by viscous action into heat). The length scale is treated in a similar manner.

The STREMR model

The model chosen in this study is based upon a two-equation k-ε rigid-lid model for two-dimensional depth-averaged incompressible flow. It was developed at the US Army Corps of Engineers Waterways Experimental Station in Vicksburg, Mississippi. It is possible to solve the Reynolds-averaged Navier–Stokes equations in their fully three dimensional form with a two-equation k-ε model. However, in streams with a relatively high width-to-depth ratio in which planform changes are dominant, it becomes both acceptable and computationally more efficient to depth-average these equations (Rodi et al., 1981) and to parameterize what three-dimensional effect would be expected. The effects of bottom friction are accounted for using the Darcy–Weisbach friction factor specified in terms of Manning's n.

Some justification of depth-averaging is required. The "dimensionality" of patterns of river channel flow will in the main be determined by the river channel topography. A number of researchers have described the importance of flow convergence and divergence in divided reaches (Smith, 1974; Church and Jones, 1982; Ashworth and Ferguson, 1986) and such flows are clearly at least two-dimensional. However, recent field observations have emphasized the importance of vertical motions in such channels. The acceptability of depth-averaging therefore needs to be considered with respect to the relative importance of these vertical motions. These may be small-scale and related to the effects of the channel boundary, and particularly micro-topographic features (e.g. Hassan and Reid, 1991; Clifford et al., 1992). Little is known about the effects that the micro-scale topography has on macro-scale flow patterns. However, Rodi *et al.* (1981) note that even in the presence of such three-dimensional effects, in many rivers the depth-averaging method may be sufficiently accurate for practical purposes. At the larger-scale, vertical motions may result from the effects of macro-topography, notably where the channel is curved and a helical secondary flow develops (e.g. Ashmore *et al.*, 1992). A correction for secondary flow is necessary in a depth-averaged model as it will result in a shear stress in excess of that associated with the standard depth-averaged model. STREMR incorporates a semi-empirical secondary flow correction which is based upon the grounds that the magnitude of this effect will be proportional to the curvature of the depth-averaged streamlines. The secondary flow correction may be switched off should the user wish. For applications such as these, depth-averaging with a correction for secondary circulation is appropriate, provided changes in bed elevation are not rapid. Bernard (pers. comm.) suggests a rule of thumb in the use of such models where the lateral scale for variation of depth and velocity should be larger than the depth itself by a factor of two.

The model is based upon a finite difference (volume) solution. A computational grid is established over the section of river channel being studied, and the depth-averaged flow equations are transformed from Cartesian co-ordinates, with arbitrary spacing ($\Delta x, \Delta y$) to curvilinear (i,j) co-ordinates with unit spacing ($\Delta i, \Delta j$) such that $\Delta i = \Delta j = 1$ between the grid lines. Each grid cell (which is square in the computational (i,j) plane) is a miniature control volume for which STREMR calculates face-centred fluxes, cell-centred velocities and cell-centred pressure. The solution imposes a rigid-lid approximation, instead of a free-surface. In this application (see below) the method of acquiring the topographic boundary condition (see below) also provided some water surface information. Non-uniform bed topography and water surface topography is accounted for by assigning a bottom elevation and a surface elevation to each grid cell. A flow rate is calculated from distributed velocity information either as input across an upstream cross-section or output from a downstream cross-section.

A modified two-equation k-ε turbulence model accounts for the horizontal transport of depth-averaged momentum by turbulence, assuming that the friction force on the channel bottom accounts for the vertical transport. This is acceptable as long as the horizontal variation of depth is gentle (Bernard, 1992). A correction to the standard k-ε model is included to counter the overprediction of turbulence energy and eddy viscosity in regions of low velocity. There is also an adjustment for the k-ε model around sidewalls where gradients in both the turbulence and mean flow are large. Although these could be resolved by the flow model, this would require an impossibly fine grid resolution. It is possible to switch off the two-equation turbulence model and replace it with a zero-equation model in which a spatially uniform eddy viscosity is specified.

The numerical solution is based upon using the input velocities to compute a mass conserving initial flow which is used as a starting condition. The code then marches forward in time, using a MacCormack predictor-corrector scheme (MacCormack, 1969; Bernard, 1992) for the momentum equation and a Euler upwind scheme (Anderson *et al.*, 1984; Bernard, 1992) for the turbulence and secondary flow equations.

Model input requirements

Model input requirements can be divided into parameters and boundary conditions; the former are specific to the model and its particular formulation, are uniformly imposed, and detailed assessment of the nature of these parameters and their effects on model performance will be considered elsewhere; the latter are specific to the particular model application and are spatially distributed.

Boundary conditions (Table 6.1) are defined by the particular problem to which the model is applied. They require detailed field measurement in their determination. They are entered into the flow model either as spatially uniform values or as random distributions. In the latter case an interpolation procedure is used to assign a value for each boundary condition to each grid cell. This is used most extensively for the water surface and bottom elevation information, but it is possible to supply distributed information on Manning's n and input velocity where available. Distributed velocity information across a cross-section is nearly always supplied as an upstream or downstream boundary condition to determine the discharge and the distribution of that discharge across the section under consideration. If distributed velocity information is available within the simulated section of stream, this may also be used to control the solution upon which the model converges.

TABLE 6.1 The boundary conditions required for flow model application

Boundary condition	Description
Input velocity	This is used to specify the velocity distribution in a cross-section at the top or bottom of the reach. It is used both to calculate the total discharge and to determine the distribution of that discharge within that particular cross-section. It is specified in terms of both x and y directions or both i and j directions
Manning's n	This controls (along with the calculated water depth from water surface and bottom elevations) the bottom frictional resistance.
Water surface elevation	This is used to "fix" the lid. Random water surface data are transformed into a water surface elevation for each grid cell through a Laplacian interpolation. Values at specified locations are treated as though they were fixed boundary conditions and a Laplace equation is solved for the remaining (unspecified) locations
Bottom elevation	This is used to "fix" the bed. Random bottom elevation data are transformed into bottom elevations for each grid cell through a Laplacian interpolation. Values at specified locations are treated as though they were fixed boundary conditions and a Laplace equation is solved for the remaining (unspecified) locations. Bottom elevation is subtracted from water surface elevation for each grid cell to determine the water depth
Eddy viscosity	This is specified when the two-equation turbulence model is switched off and the flow model uses a zero-order Boussinesq approximation with a uniform eddy viscosity value. Spatially distributed viscosity values may be specified but this is unusual due to difficulties in field determination

Field data collection and data preparation

The field site chosen for this study was a 50 m length of actively braiding proglacial stream approximately 300 m from the snout of the Haut Glacier d'Arolla in the Pennine Alps, Switzerland (Figure 6.2 and 6.3). Previous observations (1990, 1991) had revealed that this section of stream was particularly dynamic with a markedly diurnal discharge regime fluctuating between low flows of 0.75 m^3 s^{-1} and peak flows of 7.0 m^3 s^{-1}. Fieldwork was undertaken between late June and mid-August 1992. Data collection involved two main aspects; first, information on boundary conditions was required including water surface and bed topography, sedimentology and flow rate across an individual river channel section; second, distributed velocity information was needed to provide a check on model predictions. The flow model has been applied to the river channel of the 5 July, using the topographic data and velocity information collected between 10 a.m. and 1 p.m.. The discharge was constant throughout this period at 1.18 m^3 s^{-1}.

Bottom and water surface elevations

The key requirement in an attempt to model flow in this type of stream is distributed information on river-bed topography and water surface elevation. The determination of bed elevations made use of a technique that combined terrestrial analytical

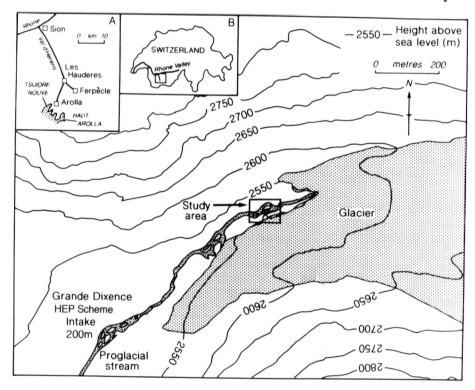

FIGURE 6.2 Location map of field-site

FIGURE 6.3 The study reach on 5 July 1993. Flow is from the left to the right-hand side of the picture

photogrammetry with rapid tacheometric survey of those areas of the bed that were not visible on the photographs. This method, and some of the results obtained, are discussed in Lane *et al.* (1994a). The acquired information consists of digital terrain models (DTMs) of the river bed, from which bed elevations can be extracted at a user-defined grid spacing and which is interpolated by STREMR to the finer computational grid spacing. These DTMs also provide information on the water surface via the water edge. This can be used to calculate the water surface elevation for each grid cell if it is assumed that the water surface is quasi-planar. Figure 6.4 shows a contour plot of the topography with annotation. The reach topography was characterized by a major diffluence at the top of the reach (A), three minor diffluences (B), a major confluence (C) and two minor confluences (D). Following the true right branch of the major diffluence, the flow moved through a markedly constrained chute before expanding into a wider section with a medial bar (E comprising a minor diffluence and minor confluence). Most of the flow travelled around the true left of the bar. The minor confluence of the medial bar was just upstream of a further minor diffluence, but the majority of the flow travelled down the left-hand branch (F) where it flowed into a minor confluence and ultimately the major confluence (C). The true left channel flows into a minor diffluence, the right-hand branch, and minor branch, flowing into F and the left-hand channel into the major confluence (C). Most of the flow travelled down the left-hand branch.

Input velocities

These were determined by measuring the depth-averaged velocity (at 0.6 of the depth for three 10-second periods) using an impellor-type current meter at a series of equally spaced verticals along a pre-established gauging section at the upstream of

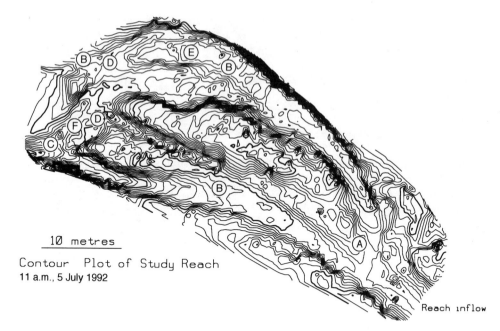

FIGURE 6.4 Contour plot of the reach used in this study

the reach (Figure 6.3). These were located such that the change in velocity between adjacent verticals was not greater than 10%. The computational grid was chosen such that the gauging section was normal to the i-axis and hence the depth-averaged velocities could be entered directly as i-direction velocities. There is a permanent gauging station 500 m downstream from the reach and the total discharge calculated from the measured velocities could be checked against this record.

Manning's n

Manning's n has no fundamental physical basis. However, various methods of empirically determining n have been developed, based upon qualitative interpretation of photographs (e.g. Barnes, 1967), qualitative assessment of various river channel properties such as channel shape and vegetation (e.g. Cowan, 1956) and quantitative assessment of river channel properties such as sedimentology (e.g. Strickler, 1923; Bray, 1979, 1982). Some distinction must be made between these different methods with reference to the scale of roughness that is being measured. Following the Cowan method for quantifying n, it is determined as the sum of individual components related to: the bed material, the level of structure associated with the bed micro-topography (the bedform contribution); the degree of variability in river cross-section shape; the relative effects of obstructions and vegetation; and the degree of variability in river planform shape. This distinction is important because at least some of the scales of roughness described above will already be incorporated into the model in terms of the bottom elevation information. Assessment of the quality of the acquired DTMs (Lane et al., in press) suggested that the bed elevation information used in this study would detect scales of morphological information between the micro-topography scale and macro-topography scale (cf. Prestegaard, 1983). Thus the value of Manning's n used in this study should reflect the contribution to the bed roughness from both the bed-material particle-size distribution and bedforms. Thus the Strickler (1923) equation that relates median particle diameter to n was used to determine an initial value of n:

$$n = 0.0151 \ D_{50}^{1/6} \tag{6.1}$$

The median particle diameter was determined from Wolman (1954) grid-by-number counts of the particle-size diameter. A 10×10 sampling mesh (10 cm separation in both planform directions) was constructed and used to define 100 sampling points. The grid was randomly located at 76 locations on exposed bar surfaces at the start of the study. This was used to determine the median particle diameter at each location and hence a value of n for each location. The result was a spatial map of Manning's n for exposed bar surfaces. Two problems arose from this. First, despite providing an assessment of the spatial variability of Manning's n, it did not provide information on subaqueous zones or the spatial variability therein. Thus, a uniform value of Manning's n was input to the model determined from the average of the samples; 0.0297 ± 0.0004. The extremely small standard error may allow some indication of the relative uniformity of Manning's n through this reach, although this reflects the relative insensitivity of Manning's n to local variation in bed sedimentology at the within-reach scale.

The second problem is with respect to the increase in n due to the presence of bed micro-topography. The percentage increase due to the contribution of micro-topography was calculated from the Darcy–Weisbach friction factor form component, in this situation being around 15–20% (Clifford et al., 1992). A value of 0.035 was therefore used in this study. The magnitude of this adjustment for micro-topographical effect, and its spatial variation, remains a key undetermined issue in applications of this kind. Research to resolve this problem through the use of two-phase photogrammetry to map micro-topography is continuing. Inclusion of greater topographical detail should reduce the importance of specification of Manning's n to that associated with skin friction which is more easily quantifiable.

Distributed velocity information

A rapid survey of the spatial patterns of flow velocity was undertaken using a Colnbrook–Valeport type electromagnetic current meter (ECM). This has a discoidal sensor head which was mounted on a wooden wading rod, such that the distance between the centre of the sensor head and the wading rod was 0.15 m and the plane of the sensor head containing the electrodes was mounted 0.05 m above and parallel to the bed. A prism was mounted on top of the wading rod allowing the precise position of the velocity measurement to be established through tacheometric survey. The advantage of the ECM is its ability to measure in two orthogonal directions, and therefore measure the magnitude of the planform velocity independently of instrument orientation. If some measure of the orientation of the sensor head is made at the time of observation, then it also becomes possible to calculate flow direction. This was possible by ensuring that the ECM x-component electrode was aligned in parallel to the supporting clamp holding the sensor head to the wading rod. It was then simply a case of orienting the wading rod such that the clamp was pointing towards the digital tacheometer (Figure 6.5). The position of the wading rod is known, as is its bearing with reference to the local co-ordinate system (β) and hence by applying a simple rotation it is possible to compute the u and v velocities from the ECM x and y velocities, although these equations require modification according to the quadrant in which the reference angle lies (Figure 6.5):

$$u = x \sin \beta + y \cos \beta \quad (6.2)$$

$$v = x \cos \beta - y \sin \quad (6.3)$$

The ECM output was sampled every 5 seconds for 25 seconds and recorded manually by reading the output from a digital datalogger.

Positions for ECM velocity measurement were randomly sampled. However, to make a comparison between STREMR depth-averaged velocities and these ECM velocities, it is either necessary to transform the ECM velocities into depth-averaged velocities or to consider only those measured velocities for which 0.05 m is approximately 0.4 of the height above the bed. The transformation is possible if it is assumed that a near-logarithmic profile exists throughout the vertical and an estimate of the bed roughness (roughness height) is specified. However, a number of authors have described marked deviations from the logarithmic profile above the bottom 20% of the flow (e.g. Petit, 1990). Intensive studies of these channels have also discovered a marked spatial variability in roughness height (Ferguson *et al.*, 1989). It was therefore decided only to make use of those ECM velocities where the flow

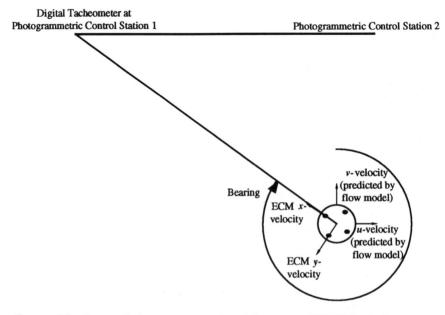

FIGURE 6.5 Geometrical arrangement of spatial patterns of EMCM velocity survey

depth was 0.125 ± 0.03 m. These locations were not evenly distributed and a map of those locations used in the assessment is provided in Figure 6.6.

Data preparation

The first stage of data preparation involved dividing the study reach into manageable parts to which the model was applied successively. There are a number of reasons for doing this, rather than considering the reach as a whole. The first is that it makes generation of the computational grid substantially easier. Second, the grid spacing required for flow predictions to be independent of the grid in use is relatively small. The required grid for the whole reach would require dimensions of approximately 600 × 200. Dividing the reach into manageable parts allowed one such part to be studied in detail in terms of its numerical behaviour, without an excessive increase in run-time. Third, the code limits the number of lines of boundary condition and parameter information input to 500. To endeavour to model the full reach in one attempt would significantly reduce the amount of boundary information that could be used in the model solution and result in greater generalization in the boundary condition information. The reach is therefore separated into subdivisions, the flow model is applied to each subdivision, and this is used to build up the total solution piece by piece. Figure 6.7 shows the river channel boundary and the jigsaw applied to the reach to model the patterns of flow on 5 July.

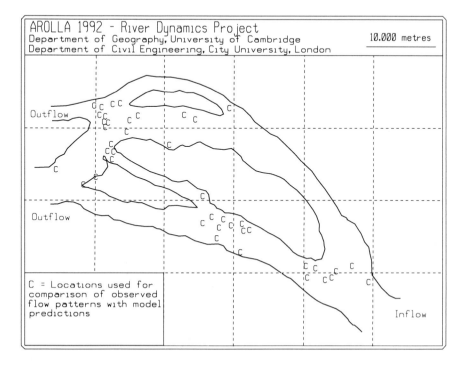

FIGURE 6.6 The spatial distribution of observed EMCM velocities used in the assessment of model predictions

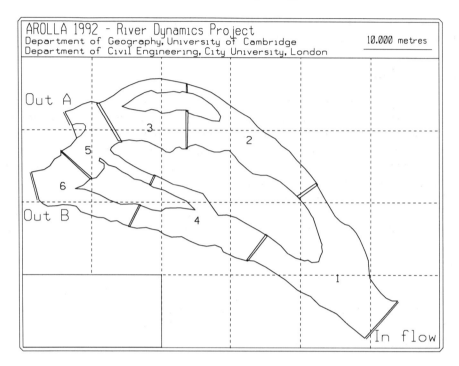

FIGURE 6.7 The division of the reach into component jigsaw pieces

The second stage of data preparation involves the generation of the computational grid for each jigsaw piece. This uses information on the water's edge obtained photogrammetrically. Each Cartesian (planform) location of the river channel boundary is given a corresponding location in the computational i–j plane. The boundary of the i–j plane does not have to be the same as the channel boundary, but grid cells in the computational plane that lie outside the channel boundary must be coded as missing segment(s). This information is normally sufficient to define an acceptable grid, but, if desired, additional information can be supplied within the Cartesian co-ordinate system to alter the precise location of grid lines. The third stage of data preparation involves entry of the required parameter and boundary condition information. Experience with application of the flow model showed that it was necessary to specify a uniform depth of 5 cm around river channel boundaries. Inflow velocity information was only available for jigsaw piece 1 and so the output velocity information provided by the flow model for this piece was used as input to pieces 2 and 4 (see Figure 6.7), the output from piece 2 was used as input to piece 3, and so on moving down the reach.

MODEL ASSESSMENT

As noted in the Introduction, before it is possible to assess the geomorphological implications of model predictions, it is important to consider model effectiveness. Detailed assessment of model stability and suitability has been undertaken and the

TABLE 6.2 Statistical comparison of observed velocities obtained from distributed ECM survey with those predicted by the flow model

	Correlation with flow model velocities	Regression analysis					No. of observations
		Slope	Standard error	Intercept	Standard error	F ratio	
Resultant velocity	0.783	0.913	0.113	0.0555	0.0628	65.1	42
u-velocity	0.717	0.799	0.121	−0.139	0.0589	43.3	42
v-velocity	0.566	0.404	0.0919	0.0551	0.0248	19.4	42

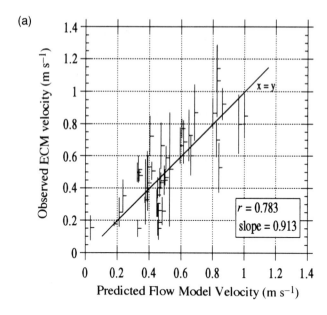

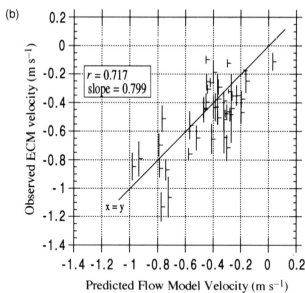

FIGURE 6.8 Plots of the STREMR predicted velocity against EMCM observed velocity for (a) resultant velocity (b) u-velocity and (c) v-velocity

model has been subjected to an extensive sensitivity analysis. These results will be presented elsewhere (Lane et al., in prep.) and suggest that the model behaviour is sensible when applied to this type of problem.

Ultimate assessment of a model of this kind must make recourse to spatially distributed measurements of internal model predictions. The plots of predicted

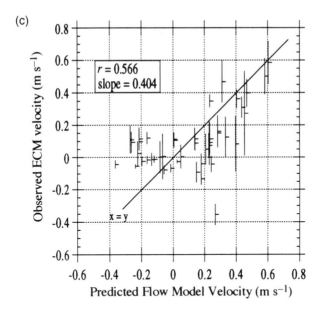

FIGURE 6.8 (*continued*)

versus observed velocities (from the field EMCM survey and with the standard errors associated with the observed velocities attached) are illustrated in Figure 6.8, with the statistical assessment being presented in Table 6.2.

There is a reasonable correlation between observed and predicted resultant velocities and u-velocities (0.783 and 0.717 respectively), but a poorer correlation between observed and predicted v-velocities (0.566). These question model effectiveness as they imply that 40% of the resultant velocities, 50% of the u-velocities and 70% of the v-velocities remain unexplained. These high levels of unexplained variance are at least in part due to the nature of the observations. The flow model makes time-integrated predictions which are being compared with a set of instantaneous (although averaged) observations, a point reinforced by the large standard errors shown on Figure 6.8. Increased confidence in model performance is obtained through consideration of the slope and intercept of the regression lines. For perfect correspondence the slope should be unity and the intercept zero. Applying t-tests (95% significance level) in the case of the resultant velocity, the slope (0.913) is not significantly different from one and the intercept is not significantly different from zero. This is not the case with respect to the u- and v-velocities, despite significant correlations. This implies that model predictions are sensible in the case of the resultant velocities and at least some of the unexplained variance in the case of resultant velocity can indeed be attributable to the problem of using sampled instantaneous velocities to assess time-averaged predictions. Some caution must also be given to the extent to which the u-velocity and v-velocity observations are correct. The magnitudes of the u- and v-velocities will be particularly sensitive to the orientation of the sensor head, which in this situation is difficult to guarantee to better than ±20° of the correct orientation (towards the digital tacheometer in this

(a)

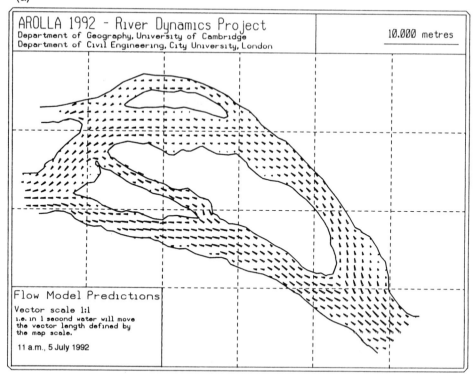

(b)

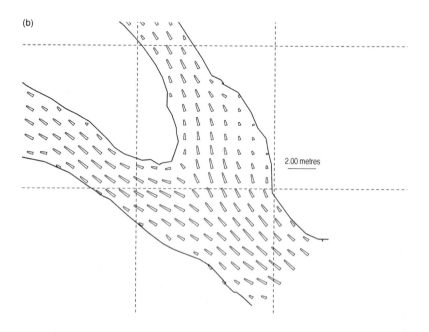

Patterns of Process and Channel Adjustment

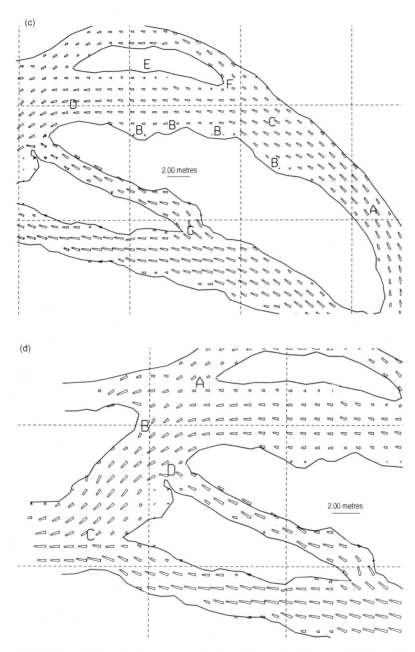

FIGURE 6.9 Plots of the velocity vectors obtained from the flow model. Long rhomboid bases are vector origins. (a) illustrates the data set for the entire reach, (b), (c) and (d) are magnified views of certain areas to clarify description provided in the text

case). This orientation problem does not affect resultant velocities as they will be independent of sensor head orientation. In summary, and recognizing the problems associated with instantaneous observations and unreliable measurement of u- and v-velocities, this field comparison is particularly encouraging if only because there has yet to be any attempt at model optimization through parameterization.

GEOMORPHOLOGICAL IMPLICATIONS

Figure 6.9a shows predicted velocity vectors for the entire study reach and Figures 6.9b, c and d, enlargements of this map to clarify interpretation. Figure 6.9a shows the major reach bifurcation. Velocities are generally high onto the bar head, despite rapidly shallowing flow, and this accords with field observations made by other researchers working in this type of environment (Ferguson, pers. comm.). This has key implications for upstream bar-growth and was observed in the field to be encouraging entrapment of coarse gravel and cobbles which both reinforced bar head stability and encouraged its later growth upstream (cf. Ashmore, 1982; Ashworth *et al.*, 1992). The magnitude of flow divergence increased to a maximum at the bar-head. Flow magnitudes were high into the left-hand branch, given the shallow flow and lateral topographical constraints.

Flow in the right-hand branch accelerates markedly downstream from the bifurcation (Figure 6.9c, location A), aided both by a steepening bedslope and lateral constriction of the channel sides. This may in part explain the strong skewing of flow towards the true right in the bifurcation as water is sucked into the branch, which accords with the observations of Whiting and Dietrich (1991). What is most notable is the relatively high velocities on *both* sides of the channel which in the subsequent period was observed to lead to rapid bank retreat, notably on the *true left* bank. This is attributable to the rapidly steepening bedslope, which encourages high velocities close to the left-hand bank. Steep longitudinal slope reduces the effects of lateral bedslope with momentum effects dominating topographic controls. Further downstream, topographic controls were more important controls on within-channel flow dynamics. A series of embayments along the true left bank (Figure 6.9c, locations labelled B), in which recirculating flow can be noted, illustrates the effect of river bank form on flow processes and future bank recession. High velocities on the bay heads concentrated bank recession activity in such locations, protecting the embayments from further retreat. The exact nature of the space–time dynamics of bank recession along this section are currently under investigation.

As the true right branch widens and becomes less steep (Figure 6.9c, location C), there is a corresponding reduction in flow velocity. This was the most dynamic section of the river channel in the subsequent few days, with the formation of a new medial bar (Figure 6.9c location D). This can be related to the effects of bank recession further upstream which supplied large amounts of fine sediment to the reach. As the left-hand bank upstream retreated (Figure 6.9c, location A), sediment supply to the left-hand side of the channel was high and led to the formation of the medial bar. This caused the thalweg to shift towards the true right bank and resulted in substantial bank retreat (Figure 6.9c, location E). The minor bifurcation in Figure 6.9c (location F) was also associated with high magnitude velocities onto the bar head but with much later flow divergence, associated with a strong lateral bedslope off the bar head into the main channel. Flow magnitudes were high and spatially varied in the channel to the true right of this bar, although this was characteristic of a very shallow riffle in which predictions were somewhat unreliable.

Flow onto the other downstream minor bifurcation shown on Figure 6.9c (location G) was also of high velocity, the flow turning into the right-hand branch extremely late because of a major protrusion on the true right branch. The flow itself into this minor channel was then rapidly skewed in the opposite direction as it responded to the orientation of this particular channel branch. This would imply the potential for much erosion in this region, given flow that is alternately converging on two river banks. However, the coarse bed and shallow depth meant that this was one of the most stable sections of the reach, not experiencing substantial change for 10 days, until upstream channel processes resulted in greater discharges through, and sediment supply to, this area.

The most complicated flow patterns are illustrated in Figure 6.9d. The medial bar (Figure 6.9d, location A) was still attached to the minor bifurcation immediately downstream of it with the result that there was little mixing of the two flows in this region. However, the marked lateral steepening of the river bed from this zone into the main channel, when combined with strong divergence late into the bifurcation in part explains the strong skewing of the flow towards the true left in this zone (Figure 6.9d, location B).

The major confluence (Figure 6.9d, location C) illustrates the combined effects of momentum and channel topography on flow patterns. The high momentum of flow in the true left channel results in a greater skewing of the flow that enters from the true right. In the right branch and upstream of the confluence, a minor confluence (Figure 6.9d, location D) supplies water to the right branch, which causes flow deflection to the left-hand branch which has to be countered further downstream, as a result of the topography on the true right bank. The result is an s-shaped flow pattern, and flow concentration in the middle of the confluence. The confluence was observed to scour rapidly during this period, and this can be attributed to the concentration of flow in this area. The position of the minor confluence (Figure 6.9d, location D) is itself delayed by the presence of a coarse boulder strip which was also responsible for a very weak circulating flow behind it. The latter was extremely deep, but because of the reverse eddy fine sediment was observed to be settling out of suspension.

CONCLUSIONS

This paper illustrates the potential of computational flow models in applications to complex natural river channels. The combination of discharge, sedimentology and rapidly acquired topographical information allows distributed velocity patterns to be modelled. Confidence can be placed in model predictions as a result of the consistency of the mathematical model, the behaviour of the numerical scheme under various imposed conditions, and because of a good correspondence between model predictions and distributed measurements of flow velocity. The latter is encouraging given that there have been no attempts to optimize model predictions.

The results presented suggest that flow patterns in a steep braided stream are a result of the interaction between momentum or inertial effects and topographical effects. Momentum effects are illustrated on bar heads, where divergence occurs relatively late, but where this delay is critically dependent upon river channel topography in the immediate locality. In the confluences, it is the combination of confluence angle and relative momentum of flows in the two channels that controls the flow pattern. This is illustrated in the major confluence where the higher momentum of the true left channel forces the right-hand flow to skew markedly. Again this cannot be divorced from topographical controls where the bar tail was maintained by cobbles and boulders, in part skewing the left-hand flow before it was joined by the right-hand channel. Flow in the branches is particularly controlled by changes in river channel width and depth, which can be responsible for marked flow acceleration and deceleration.

The results obtained in this paper show much promise for the numerical modelling of within-channel processes in dynamic alluvial channels. This potential is currently being exploited both through continued evaluation and improvement of flow model performance and through model extension to quantification of other aspects of Figure 6.1. In detail, these objectives are described in the following sections.

Continued evaluation and refinement of the flow model

It is clear from the model assessment described above that a number of refinements may be necessary to improve model predictions. This must begin with a consideration of areas where flow model predictions are particularly poor, as revealed both by the spatial variability in confidence in model predictions (Lane *et al.*, in prep.) and the extent to which there is systematic spatial variability in the difference between model predictions and observed flow patterns. It might be possible to undertake some optimization with model parameters, but given questions of the reliability of ECM observations and the problem that parameter optimization may produce the "correct" result but for the wrong reasons, such an approach must be treated with caution. More specifically, the following need investigation.

1. Grid quality; it is almost certainly the case that in certain areas of the reach studied the flow predictions are not independent of the exact nature of the grid. This needs assessment in terms of the effects of making the grid progressively finer and in terms of altering the grid shape in those areas where grid error is likely to be particularly important (around bar points and bar heads).
2. The American Society of Civil Engineers noted in 1988 that much of the development of more sophisticated turbulence closures was proceeding without due consideration of the extent to which such sophistication was required, particularly with reference to natural alluvial channels. Given the complex nature of the topographical boundary condition, it may be that the uncertainty associated with this is more important than the sophistication with which other system attributes are represented. The flow model is currently being applied with a uniform eddy viscosity to assess the effect on model prediction of the distributed velocity observations.

3. Specification of the river channel boundary conditions may be further improved. In this paper bed elevations are used which capture variation at a micro-topographical scale. At scales smaller than this, topographic effects are parameterized through the roughness coefficient. Given the uncertain physical basis for Manning's n, the problems of determining Manning's n (sedimentological sampling, relating sedimentological information to n, and correcting for the effects of micro-topography), and the overall importance of Manning's n as a control on model confidence (Lane et al., 1994b), some effort must be made either to improve the reliability of specification of n or to reduce the dependence of model performance on n. This may be achieved through a strategy that improves representation of micro-topography. Supplying such information to the flow model should reduce the importance of accurate specification of n (which would then only reflect the roughness representative of surface clasts) while at the same time allowing detailed consideration of the importance of micro-topographical bed elements in controlling larger-scale flow processes, and ultimately sediment entrainment and deposition. For this purpose, the authors are investigating photogrammetry of a stream bed in a clearwater stream. If successful, this should allow the acquisition of accurate digital terrain models of river-bed micro-topography and sedimentology, without disturbing the river bed.

4. This flow model makes use of a rigid-lid assumption. Consideration must be given to the effects of specifying the water surface as a rigid lid in this application. The importance of the water surface has been emphasized by Dietrich and Smith (1983), describing water surface super-elevation in a meander. However, the water surface was specified from accurate information at the channel margins, and generated assuming a quasi-planar surface. It is important to assess this assumption; this has been undertaken through the use of synchronous terrestrial oblique photogrammetry at a small scale to map water surface topography. Small amounts of biodegradable soap liquid were injected onto the water surface to create bubbles, which provide features for digitization with a precision of c. 5 mm. This technique has been applied to the chute-type channel (Figure 6.9c, location A) downstream of the major bifurcation that appeared from observation in the field to have maximum water surface deviation from a planar condition in the cross-stream sense. This should allow some assessment of the effects of water surface mis-specification on flow model predictions. An alternative approach is to use a free-surface model, which will have the added advantage that it allows the modelling of unsteady flow in response to a changing hydrograph, although it only has a zero-order turbulence correction.

5. Ultimately, application of a three-dimensional flow model would be of interest. The extent to which this is necessary at the scale of investigation presented here will need consideration, but certainly for modelling the flow patterns around micro-topographical bed elements, it is probably critical.

Extension of the modelling approach to the prediction of shear stress, entrainment, transport and ultimately channel change

This paper has shown that it is possible to model flow patterns numerically and such patterns can be used to make qualitative inferences about patterns of erosion and deposition. The ultimate aim is to couple transport models to the flow model and so simulate channel change. This will not be an easy task. Research into the nature of the entrainment function illustrates the complexity of sediment erosion, both in terms of the nature of the fundamental processes and their spatial variation due to interaction with spatial variable boundary conditions (bed sedimentology). The way in which this methodology could be developed is dependent upon the scale of application.

An obvious extension will begin with the prediction of the bed shear stress. This can be done either using the law-of-the-wall or using a quadratic friction law (Rodi et al., 1981) with some specification of bed roughness. This may require improved monitoring of river-bed micro-topography to reduce the uncertainty in specification of bed roughness described above. Such data will also provide critical information on bed sedimentology, which in combination with shear stress predictions may be used to model sediment entrainment and transport at the micro-scale. There is much potential to explore different entrainment functions to this effect. Given problems in specification of detailed bed sedimentology, relatively simple functions in which sedimentological effects are parameterized may be appropriate. This is reinforced when the uncertainty associated with other aspects of model formulation are considered.

In a stream of the kind described here, upstream sediment supply effects will be critical. The specification of the entrainment threshold will necessarily be simplified, but in combination with a continuity equation, may be used to predict the spatial patterns of erosion and deposition and therefore channel change. There remains much scope at this scale for stochastic and probabilistic approaches to the representation of sediment entrainment, and comparisons of bed aggradation and degradation potential with actual patterns of erosion and deposition obtained through the intercomparison of successive Digital Terrain Models.

ACKNOWLEDGEMENTS

SNL was in receipt of NERC studentship GT4/1991/AAPS/9 and acknowledges support from Fitzwilliam College, Cambridge, and the Leathersellers' Company, London. This research was supported by NERC Grant GR9/547 co-awarded to KSR and JHC. This paper was improved substantially by critical and constructive comments from Ted Hickin and an anonymous reviewer.

REFERENCES

Anderson, D. A., Tannehill, J. C. and Pletcher, R. H. 1984. *Computational Fluid Mechanics and Heat Transfer*. Hemisphere Publishing, McGraw-Hill, New York.

ASCE, 1988. Turbulence modelling of surface water flow and transport: Part V. *ASCE Journal of Hydraulic Engineering*, **114**(9), 1052–1073.

Ashmore, P. E. 1982. Laboratory modelling of gravel, braided stream morphology. *Earth Surface Processes and Landforms*, **7**, 201–225.

Ashmore, P. E. 1991. How do gravel-bed rivers braid? *Canadian Journal of Earth Sciences*, **28**, 326–341.

Ashmore, P. E., Ferguson, R. I., Prestegaard, K. L., Ashworth, P. J. and Paola, C. 1992. Secondary flow in anabranch confluences of a braided gravel-bed stream. *Earth Surface Processes Land and Forms*, **17**, 299–331.

Ashworth, P. J. and Ferguson, R. I. 1986. Interrelationships of channel processes, changes and sediments in a proglacial braided river. *Geografiska Annaler*, **68A**, 361–371.

Ashworth, P. J., Ferguson, R. I. and Powell, M. D., 1992. Bedload transport and sorting in braided channels. In: Billi, P. Hey, R. D., Thorne, C. R. and Tacconi, P. (eds), *Dynamics of Gravel-bed Rivers*, Wiley, Chichester, 497–515.

Barnes, H. H. 1967. *Roughness Characteristics of Natural Channels*. Water Supply Paper 1849, USGS, Washington, DC.

Bernard, R. S. 1992. STREMR: Numerical model for depth-averaged incompressible flow. Tech. Rpt REMR-HY-105, Army Corps Engineers Waterways Experimental Research Station, Vicksburg, USA.

Boussinesq, J., 1877. *Essai sur la théorie des eaux courantes*. Mémoires présentés pars divers savants à l'Académie des Sciences, Paris.

Bray, D. I. 1979. Estimating average velocity for gravel-bed rivers. *ASCE Journal Hydraulics Division*, **105**(9), 1103–1122.

Bray, D. I. 1982. Flow resistance in gravel-bed rivers. In: Hey, R. D., Bathurst, J. C. and Thorne, C. R. (eds), *Gravel-bed Rivers*, Wiley, Chichester, pp. 109–133.

Carson, M. A. 1971. *The Mechanics of Erosion*, Pion, London.

Church, M. and Jones, D. 1982. Channel bars in braided rivers. In: Hey, R. D., Bathurst, J. C. and Thorne, C. R. (eds), *Gravel-bed Rivers*, Wiley, Chichester, pp. 291–339.

Clifford, N. J., Robert, A. and Richards, K. S. 1992. Estimation of flow resistance in gravel-bed rivers: a physical explanation of the multiplier of roughness length. *Earth Surface Processes Land and Forms*, **17**, 111–126.

Cowan, M. W. 1956. Estimating hydraulic roughness coefficients. *Agricultural Engineering*, **7**, 473–475.

Deissler, R. G. 1977. Turbulence processes and simple closure schemes. In: Frost, W. and Moulden, T. H. (eds), *Handbook of Turbulence, Vol. 1, Fundamentals and Applications*, Plenum, New York, 165–186.

Dietrich, W. E. and Smith, J. D. 1983. Influence of the point bar on flow through curved channels. *Water Resources Research*, **19**, 1173–1192.

Dietrich, W. E. and Smith, J. D. 1984. Bedload transport in a river meander. *Wat. Res. Res.*, **20**, 1355–1380.

Emmons, H. W. 1970. *Annual Review of Fluid Mechanics*, **2**, 15–36.

Ferguson, R. I., Prestegaard, K. and Ashworth, P. J. 1989. Influence of sand on hydraulics and gravel transport in a braided gravel-bed river. *Water Resources Research*, **25**, 635–643.

Hassan, M. and Reid, L. 1991. The influence of microform bed roughness elements on flow and sediment transport in gravel-bed rivers. *Earth Surfaces Processes and Landform*, **15**, 739–750.

Howes, S. and Anderson, M. G. 1988. Computer simulation in geomorphology. In: Anderson, M. G. (ed.), *Modelling Geomorphological Systems*, Wiley, Chichester, pp. 421–440.

Lane, S. N., Chandler, J. H. and Richards, K. S. (1994a). Developments in monitoring and terrain modelling small-scale river bed topography. *Earth Surface Processes and Landforms*, **19**, 349–68.

Lane, S. N., Richards, K. S. and Chandler, J. H. (1994b). Distributed sensitivity analysis in environmental modelling. *Proceedings of the Royal Society Series A*, **447**, 49–63.

Lane, S. N., Richards, K. S. and Chandler, J. H. (in preparation). Use of a depth-averaged flow model with a second order turbulence closure to predict flow processes associated with complex river bed topography.

Leopold, L. B., Wolman, M. G. and Miller, J. P. 1964. *Fluvial Processes in Geomorphology*. Freeman, San Francisco.

MacCormack, R. W. 1969. The effect of viscosity in hypervelocity impact cratering. *AIAA Paper* **69-534**, Cincinatti, Ohio.

McCuen, R. H. 1973. The role of sensitivity analysis in hydrologic modelling. *Journal of Hydrology*, **18**, 37–53.

Naden, P. and Brayshaw, A. C. 1987. Bedforms in gravel-bed rivers. In: *River Channels: Environment and Process*, IBG Spec. Pub., **18**, 249–271.

Olson, R. M. and Wright, S. J. 1990. *Essential of Engineering Fluid Mechanics*. Harper Row, New York, 638 pp.

Petit, F. 1990. Evaluation of grain shear stresses required to initiate movement of particles in natural rivers. *Earth Surface Processes and Landforms*, **15**, 135–148.

Pickup, G. 1976. Adjustment of stream-channel shape to hydrologic regime. *Journal of Hydrology*, **30**, 365–373.

Prandtl, L. 1925. Liber die ausgebildete turbulenz. *Z. Angew. Math. Mech.*, **5**, 136–139.

Prestegaard, K. K. 1983. Bar resistance in gravel-bed streams in bankfull stage. *Wat. Res. Res.*, **19**, 472–476.

Reynolds, O., 1895. On the dynamical theory of incompressible viscous fluids and the determination of the criterion. *Philosophical Transactions of the Royal Society*, **186A**, 123–64.

Richards, K. S. 1988. Fluvial geomorphology. *Progress in Physical Geography*, **12**, 435–456.

Rodi, W. 1980. *Turbulence Models and their Application in Hydraulics*, IAHR, Delft, 104 pp.

Rodi, W., Pavlovic, R. N. and Srivatsa, S. K. 1981. Prediction of flow and pollutant spreading in rivers. In: Fischer, H. B. (ed.), *Transport Models for Inland and Coastal Waters*, Academic Press, New York, 63–111.

Smith, N. D. 1974. Sedimentology and bar formation in the Upper Kicking Horse River, a braided outwash stream. *Journal of Geology*, **82**, 205–223.

Strickler, A. 1923. Beitrage zur Frage der Geschwindigheitsformel un der Rauhigkeitszahlen fur Strome, Kanale und Geschlossene Leitungen. *Mitteilungen des Eidgenossischer Amtes fur Wasserwirtschaft*, Bern, Switzerland, 16 pp.

Whiting, P. E. and Dietrich, W. D. 1991. Convective accelerations and boundary shear stress over a channel bar. *Wat. Res. Res.*, **27**, 783–796.

Wolman, M. G. 1954. A method of sampling coarse river bed material. *Transactions American Geophysical Union*, **35**(6), 951–956.

7

Modelling Contemporary Overbank Sedimentation on Floodplains: Some Preliminary Results

A. P. NICHOLAS AND D. E. WALLING

Department of Geography, University of Exeter, UK

ABSTRACT

A mathematical model which aims to predict rates and patterns of floodplain sedimentation is presented. The model has two main elements: first, a hydraulic component which determines floodplain inundation and flow patterns and, second, a sediment transport component which models the dispersion of suspended-sediment particles away from the main channel, together with their deposition on, and remobilization from, the floodplain surface. Field monitoring of a 1 km reach of the River Culm, Devon, UK, has provided data for the model's input requirements in the form of stage hydrographs and suspended-sediment load data for individual floods, together with flood frequency information recorded over longer time periods. The model predicts depths of aggradation and scour across the floodplain and the *effective* and *ultimate* grain-size distributions of sediment within the water column and of material deposited upon the floodplain. These predictions have been tested for individual flood events with the aid of samples of deposited floodplain material, and for time periods of approximately 30 years using measurements of the caesium-137 inventories of floodplain sediments.

INTRODUCTION

The floodplains of lowland rivers are regions of considerable geomorphological, environmental and economic importance. Despite substantial efforts to document and interpret the topography and alluvial stratigraphy of these areas (cf. Lewin, 1978; Brown and Keough, 1992), the link between form and process remains relatively poorly understood. The sediments of which these alluvial units are composed may be effectively divided into lateral accretionary deposits, associated with aggradation and migration of the main channel, and vertical overbank sediments, resulting from the deposition of fine particles in floodplain regions during times of peak flow (Allen, 1965). Although channel migration may be significant over long time periods or where bank erodibility is high, this study confines itself to an examination of contemporary overbank sedimentation. It is thus of particular relevance to rivers where channel banks and sediment load are composed of fine cohesive material and where overbank flooding is relatively frequent.

At the basin scale, field investigators have shown average rates of overbank sedimentation to be typically of the order of 1 to 10 mm per year (Shotton, 1978; Trimble, 1981). As such they represent a significant proportion of annual river suspended-sediment loads and an important sediment sink (cf. Walling et al., 1986). Concealed within such mean values are important temporal and spatial variations resulting from the effects of local topography upon flow hydraulics and sediment availability, and the magnitude and frequency of overbank flood events. While attempts have been made to document these variations empirically (cf. Lambert and Walling, 1987), little progress has been made in linking hydraulic and sedimentological variables with the aid of quantitative predictive schemes. James (1985) and Pizzuto (1987) have modelled the turbulent diffusion of fine sediment across floodplains resulting from the lateral momentum exchange mechanism identified by flume workers (cf. Wright and Carstens, 1970; Yen and Overton, 1973). Although the results of these models have been supported by both laboratory and fieldwork, because they address transport and deposition processes in cross-section they are difficult to apply where channel sinuosity is high. Howard (1992) presents a model of channel migration and floodplain sedimentation which includes a simple treatment of overbank deposits in two lateral dimensions, but this model is directed towards long-term prediction of floodplain development, rather than contemporary time-scales. In order to simulate sedimentation processes at the event scale, models must be capable of predicting detailed hydraulic patterns resulting from specific flood discharges and relating these to sediment transport processes. Finite difference and finite element techniques have provided a basis for simulating flood flow patterns over complex floodplain and river channel environments (e.g. Samuels, 1985; Urban and Zielke, 1985), but it is only recently that such approaches have been adopted for geomorphological purposes (Gee et al., 1990), and even in these instances problems of scale and topographic representation remain to be overcome (Bates et al., 1992).

In this paper a mathematical model which aims to predict rates and patterns of floodplain sedimentation over a range of temporal scales is presented. Hydraulic patterns are modelled using simple "water surface functions", which enable very rapid approximations of flow conditions to be derived, even in complex topographic situations. This allows emphasis, in terms of computer time, to be placed upon the model's sedimentation component, and permits examination of time periods beyond the event scale.

MODELLING FLOODPLAIN SEDIMENTATION

The numerical solution of the two-dimensional depth-averaged form of the Navier–Stokes equations has been widely used for modelling hydraulic patterns on floodplains. However, this approach places considerable demands on computer resources and may be prone to difficulties where topographic variability is significant. To overcome this limitation a simplified hydraulic scheme has been developed which aims to model the overall structure of the flow, whilst minimizing computational requirements.

It has been suggested that flow patterns in overbank situations are largely a product of the dominant topographic features. Thus, water within the river flows parallel to the channel banks, whilst water on the floodplain itself moves parallel to the valley walls (Yen and Yen, 1983; Kiely, 1990). These general patterns are adjusted at the channel/floodplain interface as a result of flow interactions caused by steep depth and roughness differentials and differences in main channel and floodplain flow directions (Ervine and Ellis, 1987). It is argued here that, although hydraulic patterns will vary between individual floods, the overall flow structure identified above will remain constant, for a given reach, until significant changes in channel and floodplain morphology occur. A two-dimensional finite difference grid has been employed by this model to represent the channel/floodplain environment, and the dominant hydraulic patterns resulting from the topographic features within this grid are simulated using what is termed a "water surface function". This function defines the water surface at each node in the solution domain as a percentage of the difference between the water levels at the upstream and downstream domain boundaries:

$$W_N = \frac{S_U - h_N - z_N}{S_U - S_D} \tag{7.1}$$

where W_N is the water surface function value, S_U and S_D are the upstream and downstream water levels, h_N is the depth of flow and z_N is the bed elevation (N indicates a nodal variable). Water surface function values are expressed as percentages so that, having determined values of W_N for each node in the solution domain, nodal depths may be calculated for any given pair of upstream and downstream water levels. Water surface function values are determined by solving the depth-averaged continuity of mass equation (7.2), and substituting the predicted depth values back into equation (7.1):

$$\frac{\partial(uh)}{\partial x} + \frac{\partial(vh)}{\partial y} = 0 \tag{7.2}$$

where u and v are velocity components in the x and y directions respectively.

Velocity terms are approximated using a modified version of the Manning equation:

$$u = \frac{h^{0.67} S_x^{0.5}}{n} \tag{7.3}$$

$$v = \frac{h^{0.67} S_y^{0.5}}{n} \tag{7.4}$$

where n is the Manning roughness coefficient and S_x and S_y are the water surface slopes in the x and y directions. Use of the Manning equation to calculate velocity allows equation (7.2) to be written in terms of a single variable (h), thus considerably simplifying the procedure for determining values of W_N. Equation (7.2) is solved for each node in the solution domain using the Newton Raphson iterative method. In doing this, upstream and downstream boundary water levels must be specified. This is achieved by calculating the discharge entering the reach at the upstream boundary, and determining the downstream water level required to yield the same flow rate. This involves slicing the upstream and downstream cross-sections into a number of segments and applying the Manning equation to each. Such a method implies a single-valued stage/discharge relationship and assumes no loss of discharge between the two sections. However, it is suggested that this approach is reasonable when applied to relatively short reaches for which stage data are available for verification purposes. Because the degree of interaction between channel and floodplain flows varies with the relative depths in these zones, water surface function values are calculated for a range of upstream water levels at 10 cm increments. These numerically derived flow surfaces are assumed to capture the dominant hydraulic patterns in the reach, resulting from its topographic structure. They are subsequently used to determine inundation patterns for given upstream water levels using the following procedure:

(i) The downstream water level is determined using the discharge estimation method described above.
(ii) Water surface function values are calculated for the given upstream stage as linear combinations of the numerically derived functions.
(iii) Nodal flow depths are established by inserting the interpolated surface function values into a rearranged version of equation (7.1):

$$h_N = S_U - (S_U - S_D) W_N - z_N \tag{7.5}$$

(iv) Depth values are adjusted using a ponding algorithm to allow for backwater effects. Such a procedure becomes necessary in some areas, because equation (7.2) is satisfied completely only for the numerically derived water surface functions and not for the interpolated function values used in equation (7.5).

Having calculated the depth of inundation at each node, the distribution of suspended sediment across the floodplain is determined by solving the time-independent, depth-averaged mass balance equation for suspended sediment under the influence of convective and diffusive transport in two dimensions:

$$\frac{\partial}{\partial x}\left(\varepsilon_x h \frac{\partial C}{\partial x}\right) + \frac{\partial}{\partial y}\left(\varepsilon_y h \frac{\partial C}{\partial y}\right) - uh \frac{\partial C}{\partial x} - vh \frac{\partial C}{\partial y} = DR - ER \tag{7.6}$$

where C is the vertically integrated suspended-sediment concentration, ε_x and ε_y are lateral eddy diffusivity coefficients, and DR and ER are deposition and erosion terms.

In order to solve equation (7.6), the deposition and erosion terms and the eddy diffusivity coefficients must be written as functions of the hydraulic variables and sediment concentrations. According to Parker (1978) the deposition term can be given by:

$$DR = \frac{V_s^2}{\varepsilon_z} Ch \qquad (7.7)$$

where V_s is the particle fall velocity and ε_z is the vertical diffusivity. Examples of suitable erosion functions are more difficult to find in the literature. In this study a term has been chosen so as to express erosion as a function of the flow conditions, whilst maintaining the dimensions $(L\,T^{-1})$ required by equation (7.6):

$$ER = \left(\frac{u}{u_{cr}} - 1\right)^3 u_* C_{fp} \qquad u > u_{cr} \qquad (7.8)$$
$$ER = 0 \qquad u \leq u_{cr}$$

where u, u_{cr} and u_* are the predicted flow velocity, critical velocity for scour and shear velocity respectively, and where C_{fp} is the concentration of sediment of a given size class in the surface material of the floodplain. The value of u_{cr} has been defined arbitrarily as 0.75 m s^{-1}, so as to broadly limit erosion to regions of the floodplain identified during the fieldwork as areas of potential degradation. Values of C_{fp} are determined for each floodplain node from an initial model run. The shear velocity is given by:

$$u_* = (ghS_w)^{0.5} \qquad (7.9)$$

where S_w is the water surface slope. Eddy diffusivity coefficients are given by:

$$\varepsilon = \lambda u_* h \qquad (7.10)$$

Previous workers (Parker, 1978; Fisher et al., 1979) have suggested values of 0.077 and 0.13 for vertical and lateral dimensionless eddy diffusivity coefficients (λ). These figures are, however, for open channels and their applicability to overbank flows, for which few data are available, is uncertain. Shiono and Knight (1991) document results from the SERC Flood Channel Facility which indicate increases in eddy diffusivity coefficients of several orders of magnitude at the channel/floodplain interface. They show that the increase in eddy diffusivity is a function of the ratio of floodplain to main channel depth. A similar function is used here to determine λ as:

$$\lambda = k(1 - \Delta h/2\bar{h})^{-4} \qquad (7.11)$$

where k is an arbitrary constant (chosen here as 0.077 and 0.13 for vertical and lateral eddy diffusivities respectively, so as to give mean values of λ of the correct order of magnitude) and $\Delta h/2\bar{h}$ describes the relative change in flow depth between two adjacent nodes.

Having specified the elements of equation (7.6) in terms of the hydraulic variables and sediment concentrations, it can be solved at each point in the finite difference grid. In order to do this a number of boundary conditions must be specified. First, sediment concentrations in the main channel are held at a fixed value equal to that measured at the upstream domain boundary (field measurements have shown this to be a reasonable approximation). Second, the sediment concentrations across the upstream domain boundary are determined by solving the one-dimensional form of equation (7.6), which includes only lateral transport of suspended sediment away from the main channel. Finally, at the edges of the flow domain the following boundary conditions must be satisfied:

$$\frac{\partial C}{\partial x} = 0 \qquad (7.12)$$

$$\frac{\partial C}{\partial y} = 0 \qquad (7.13)$$

Thus, the boundary condition is that the concentration gradient is zero at the solution boundary, therefore no sediment travels across the junction between wet and dry nodes. The differential equation (7.6) is written as a linear difference equation and again solved using Newton Raphson iteration. This procedure is carried out once for each of the grain-size classes employed by the model.

The difficulty of obtaining representative hydraulic and sedimentological data means that the numerical model described above is not easy to test rigorously. Despite this limitation, an attempt has been made to compare model predictions of hydraulic patterns, deposition rates and the grain-size composition of deposited sediment with data collected from a detailed field study of the River Culm, Devon, UK.

THE STUDY AREA

The River Culm is a tributary of the River Exe, joining the latter 3 km north of Exeter (Figure 7.1). In its lower reaches the Culm flows in a meandering gravel-bed channel which is approximately 12 m wide with 1.5 m high banks composed of fine-grained cohesive material. Overbank flooding is relatively frequent during the winter months with substantial inundation of the floodplain occurring on about seven occasions each year.

A 40 ha study site was selected on the floodplain of the River Culm at Rewe. This includes a 900 m length of main channel. At this location the floodplain reaches widths of up to 600 m and contains a variety of topographic features such as major and minor drainage ditches, natural levees, depressions and bank breaches. The main channel also exhibits variation in its natural geometry and the reach includes a number of tight meander bends, abandoned channels and backwater areas. Figure 7.2 shows the topography of the study reach. This representation is based on an interpolation of the finite difference grid used by the model, which was derived from an intensive survey of the floodplain which included measurements of channel cross-sections at downstream intervals of less than 10 m. At the upstream end of the reach the main channel is relatively straight and located against the side of the steep hillslope which delimits the western edge of the floodplain. The eastern floodplain boundary is represented by a more gentle increase in elevation marking the transition from the rough pasture of the floodplain to the arable land above. This wide valley floor is crossed by a number of minor drainage ditches (which are typically 20 to

Contemporary Overbank Sedimentation on Floodplains 137

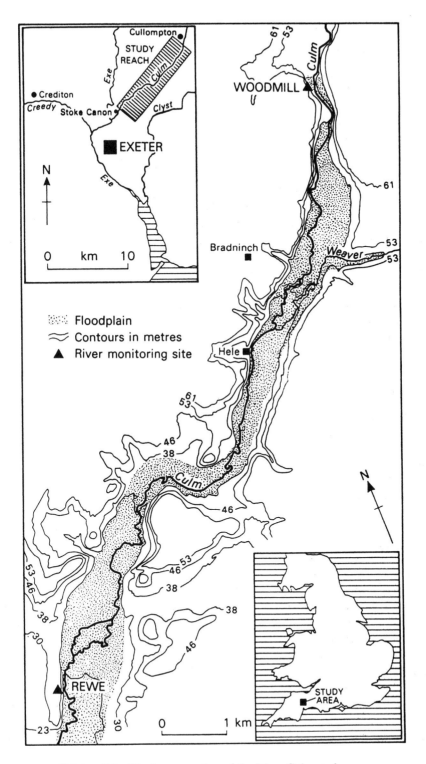

FIGURE 7.1 The lower reaches of the River Culm study area

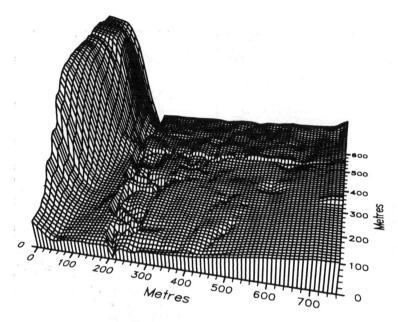

FIGURE 7.2 The floodplain topography represented by the model's finite difference grid (vertical exaggeration 1:25)

30 cm deep) and, more significantly, by a major drainage ditch which runs approximately east–west across the whole floodplain 200 m south of the site's upstream boundary.

PREDICTION OF HYDRAULIC PATTERNS

The first step in applying the model's hydraulic component is the determination of the relationship between upstream and downstream water levels. Initially, discharges were calculated using mean channel and valley slopes of 0.000864 and 0.001295 respectively for both upstream and downstream sections. Estimates of Manning roughness coefficients were taken from Chow (1959) as 0.05 for floodplain areas and 0.04 for the main channel. Figure 7.3 shows the predicted relationship between upstream and downstream stage (A) compared with the actual relationship determined from stage measurements at the boundaries of the study area. The general shape of the measured curve is replicated by the model prediction. However, for a given upstream stage, the downstream stage is underestimated by approximately 30 cm. In an attempt to improve the fit, valley and channel slopes were calculated separately for the two sections using 10 m trends in the survey data. This procedure yielded channel slopes of 0.029 and 0.019 and valley slopes of 0.00393 and 0.000328 for the upstream and downstream sections respectively. The significant increase in the channel slopes is a result of the fact that both upstream and downstream reach boundaries are located upon steep riffle sites. The stage/stage relationship derived using these slopes (B) is also shown in Figure 7.3, and it is this

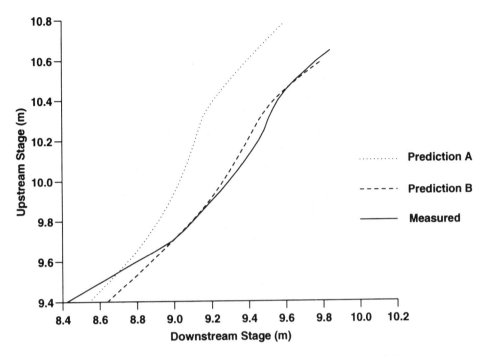

FIGURE 7.3 Predicted and measured relationships between upstream and downstream stage for the study reach

predicted curve which has been used in all subsequent model runs. The calculation procedure is clearly sensitive to the values of valley and channel gradient employed. It thus appears wise to apply this method only where both topographic and hydraulic data are available for verification purposes.

Having derived the stage/stage relationship, water surface function values were calculated for the solution domain using upstream stage increments of 10 cm. Figure 7.4 shows a typical example of such a surface for an upstream stage of 10.6 m (approximately equivalent to a one in four year flood event). The area of predicted floodwater inundation in this diagram is marked by lines joining points of equal water level, with the drop in water level between adjacent lines representing 2% of that between the upstream and downstream boundaries of the reach. The dominant features of the predicted flow structure are, as suggested earlier, floodplain water moving parallel to the valley walls and main channel water moving parallel to the channel banks, with a gentle transition between these directions where the flows interact. These patterns are most noticeable in the downstream half of the study reach where two tight meander bends cause significant twisting of the flow lines. Although rigorous testing of these predictions has not been possible, due to lack of hydraulic data, the overall flow structure identified in the model predictions compares well with the concepts put forward earlier. In addition to this, the model appears to be capable of simulating changes in flow direction witnessed in smaller scale topographic features such as the minor drainage ditches which cross some areas

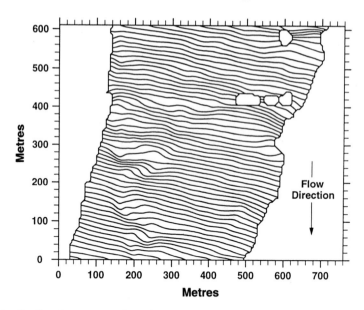

FIGURE 7.4 Predicted water surface function for an upstream stage of 10.6 m. Water surface contours are at intervals of approximately 1.5 cm

of the floodplain. The resulting divergences from the dominant downvalley flow direction are evident in Figure 7.4, although they are difficult to identify due to the essentially superficial nature of these topographic features relative to the main river channel.

PREDICTION OF SEDIMENTATION PATTERNS

Comparison of model predictions with measured sedimentation patterns has been undertaken at two time-scales. For individual floods this was achieved with the aid of samples of deposited floodplain material (samples A1 to D1 below), while caesium-137 measurements of floodplain sediment cores (samples A2 to D2 below) were employed to estimate mean deposition rates over the past 30 years. The locations of these sampling points within the study reach are shown in Figure 7.5.

Figure 7.6 shows the record of river stage and suspended-sediment concentration derived from monitoring equipment at the upstream boundary of the study area for the flood event which occurred on 9–10 April 1993. Additional information regarding the typical grain-size composition of the suspended load transported through the study reach, were obtained for a number of flood events using a custom-built water elutriation system as described by Walling and Woodward (1993). This apparatus pumps water from the river through a series of sedimentation tubes which separate particles into five *effective* size classes according to their settling characteristics, i.e <8 µm, 8–16 µm, 16–32 µm, 32–63 µm and >63 µm (sizes represent equivalent spherical diameters for particles with a density of 2.65 g cm^{-3}). Over the period from September 1992 to January 1993 the system was operated on

Contemporary Overbank Sedimentation on Floodplains 141

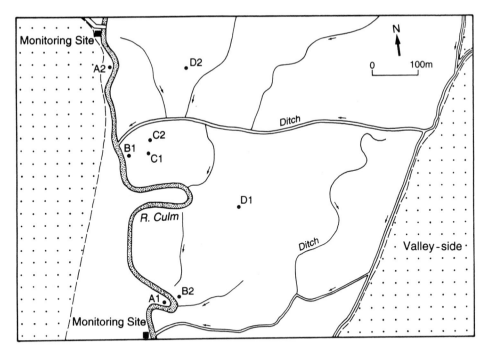

FIGURE 7.5 Sampling locations for flood deposits A1 to D1 and floodplain sediment cores A2 to D2

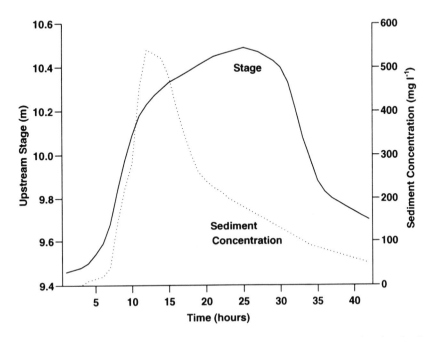

FIGURE 7.6 Monitored records of stage and suspended-sediment concentration for the flood event of 9–10 April 1993

eight occasions, with each run yielding similar *effective* size distributions. In all cases the coarser fractions represented a decreasing proportion of the load. Mean percentages for each class determined over this period were used in all model runs described below. These are <8 µm: 53%, 8–16 µm: 25%, 16–32 µm: 13%, 32–63 µm: 7% and >63 µm: 2%. These classes are designated *effective* size classes because they relate to the size characteristics of the suspended sediment *in situ* which includes both discrete and composite particles. The term *ultimate* is used hereafter to refer to the particle-size distribution of the discrete primary grains of the mineral fraction of the suspended-sediment load. *Ultimate* size distributions were determined for sediment from each of the *effective* classes for the purpose of model parameterization. This was achieved using a Malvern MasterSizer laser diffraction particle-size analyser, following the removal of organic matter and chemical dispersion of aggregates. The measurements of stage, suspended-sediment concentration and particle-size characteristics described above were used to provide data for the model's upstream boundary conditions and to test the model at the event scale. The flood record represented by these data for the storm event of 9–10 April 1993 was divided into 42 equal time periods of one hour duration, and patterns of inundation, sediment dispersion and deposition were determined for each. In this way, predictions of deposition rates and particle-size distributions of deposits were made for each node in the solution domain.

Figure 7.7 shows the predicted distribution of suspended-sediment concentrations across the study reach for an upstream water level of 10.49 m and a main channel suspended-sediment concentration of 180 mg l^{-1} (conditions experienced at the peak of the event of 9 April 1993). Solid lines in this figure join points of equal sediment concentration (adjacent lines are separated by a difference of 20 mg l^{-1}), while the

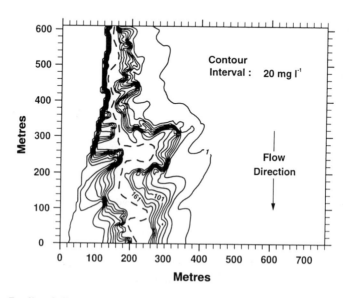

FIGURE 7.7 Predicted distribution of suspended sediment for conditions at the peak of the flood event of 9–10 April 1993. Sediment concentrations are at intervals of 20 mg l^{-1}

dashed line marks the location of the main channel. Sediment concentration decreases with distance from the channel at rates which are strongly dependent upon floodplain topography and main channel sinuosity. Where bank breaches occur and where ditches run away from the main channel into the floodplain, sediment spreads further due to the greater flow velocities and turbulence in these zones. Conversely, shallow flow over levees and other areas of high ground is characterized by high relative rates of deposition, and hence steep concentration gradients or zones of anomalously low concentration. In addition to these patterns, it is noticeable that in the downstream half of the study reach, where main channel sinuosity is higher and meander belt width greater, suspended sediment is able to spread further due to enhanced mixing of channel and floodplain waters resulting from greater interaction between these flow regions.

Measurements of the amount of deposition at a number of points on the floodplain were derived for the flood of 9–10 April 1993 using sedimentation traps similar to those employed by Lambert and Walling (1987). These consisted of small pieces of astroturf (700 cm^2) which were fixed to the floodplain surface with steel pins. Traps were located in groups of nine with each group covering an area of approximately 1 m^2, and giving a mean level of sedimentation and a measure of variability. Following the recession of floodwaters, traps were removed and returned to the laboratory where the sediment was dried and weighed prior to determination of its *ultimate* grain-size characteristics. The latter was again carried out using a Malvern MasterSizer laser diffraction particle-size analyser to examine the chemically dispersed mineral fraction of the deposited sediment. Table 7.1 gives the details of information gathered from four sets of sedimentation traps for this flood event, along with predicted levels of deposition for these sites derived by the model. Both measured and predicted data broadly indicate decreasing amounts of deposition with distance from the channel, although levels may be increased in areas such as depressions which are susceptible to ponding. This overall picture is further elucidated by examining the grain-size composition of these deposits. Figure 7.8 shows the particle-size distributions of flood deposits A1, B1 and C1. No distribution is shown for deposit D1 because insufficient sediment was retrieved from the sedimentation traps to allow particle-size analysis. The three bars shown in Figure 7.8 for each of the size classes relate to, from right to left, the predicted *effective* size distribution, the predicted *ultimate* size distribution and the measured *ultimate*

TABLE 7.1 Measured and predicted deposition levels for the flood event of 9–10 April 1993

Deposit	A1	B1	C1	D1
Trap location	Point Bar	Levee Top	Depression	Floodplain
Distance from channel	5 m	5 m	45 m	100 m
Measured deposition	166 g m^{-2}	721 g m^{-2}	256 g m^{-2}	8 g m^{-2}
Coefficient of variation	20.9%	12.2%	10.0%	76.7%
Predicted deposition	219 g m^{-2}	693 g m^{-2}	171 g m^{-2}	3 g m^{-2}

144   *River Geomorphology*

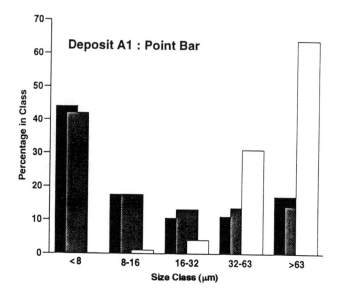

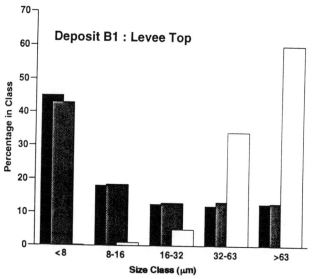

FIGURE 7.8 Predicted and measured particle-size distributions for three deposits from the flood event of 9–10 April 1993. Black bars represent measured *ultimate* grain-size distributions, grey bars represent predicted *ultimate* grain-size distributions and white bars represent predicted *effective* grain-size distributions

size distribution. All the *effective* distributions are dominated by material >32 μm with less than 2% of sediment being <16 μm. Conversely between 40% and 55% of both the predicted and measured *ultimate* material is found in the <8 μm size class with the remaining material being spread fairly evenly between the other four size classes. Although all three deposits exhibit these general patterns the *ultimate* and

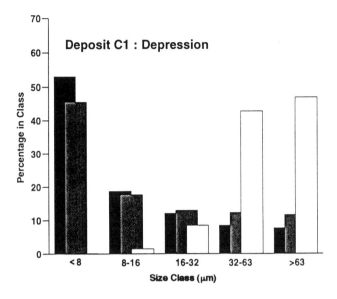

FIGURE 7.8 (*continued*)

effective distributions of deposit C1 are noticeably finer than those of either A1 or B1. As deposit C1 is located further from the river than either A1 or B1 these results fit well with the concept of decreasing mean grain size with distance from the channel supported by James (1985) and Pizzuto (1987). The most significant aspect of these data may be the difference between the degree of fining for *ultimate* and *effective* distributions. A recent study of flood deposits on the River Severn, UK, by Marriott (1992) indicated that samples of deposited material could be broadly divided into two classes on the basis of textural relationships. Within 20 m of the main channel deposits consisted largely of sand-sized grains. At greater distances sand made up only a small percentage of material while proportions of silt and clay remained roughly constant showing no marked trend of decreasing grain size with distance. Although the suspended load of the Culm is comparatively free of sand, the ultimate size distributions shown in Figure 7.8 are otherwise in close agreement with Marriott's findings. They suggest that the grain-size distributions of the dispersed mineral fraction of deposits may show only weak evidence of fining with distance from the main channel. However, Figure 7.8 also shows that the predicted *effective* grain-size distributions, which represent the *in situ* physical characteristics of the suspended sediment, exhibit more dramatic reductions in coarse size fractions and thus mean particle size. These results have important implications for the role of sediment aggregation in transport and deposition processes.

Predictions of mean annual deposition rates within the study reach, as a first approximation, were derived from the model with the aid of flow and sediment data representative of conditions over longer time periods. Upstream water levels were specified using a flow duration curve based upon data collected over the period from 1963 to 1992 by the National Rivers Authority. Suspended-sediment concentrations

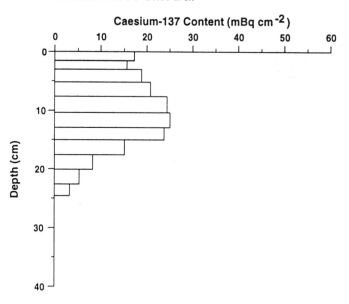
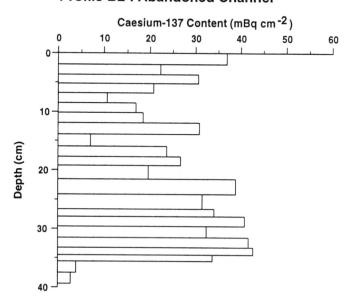

FIGURE 7.9 Caesium-137 profiles for sediment cores taken at four floodplain sites

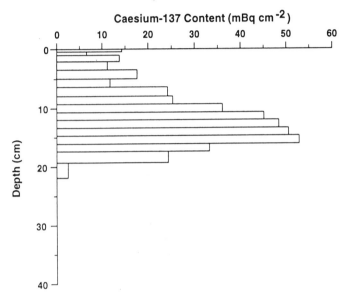
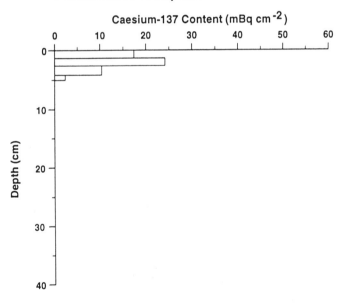

FIGURE 7.9 (continued)

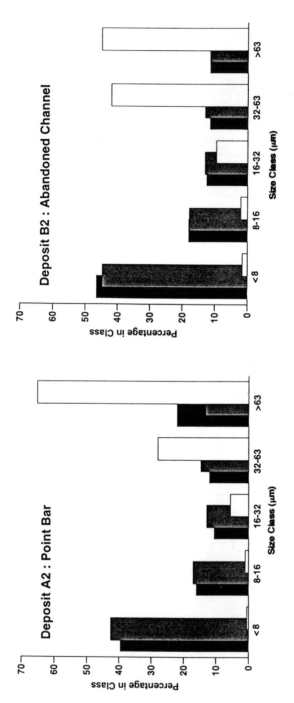

FIGURE 7.10 Predicted and measured particle-size distributions for samples of surface material taken at four floodplain sites. Black bars represent measured *ultimate* grain-size distributions, grey bars represent predicted *ultimate* grain-size distributions and white bars represent predicted *effective* grain-size distribution

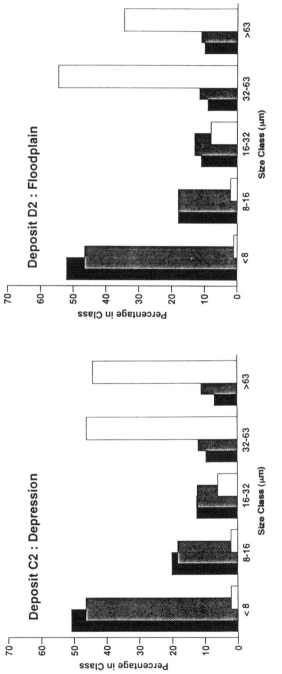

FIGURE 7.10 (continued)

were determined using the following relationship established from data collected at the upstream site boundary:

$$C = 234.7 S_U - 2241.5 \qquad (7.14)$$

where C is the average of all suspended-sediment concentrations recorded during the monitoring period (January 1992 to April 1993), which corresponded with an upstream water level of S_U. Having thus fulfilled the boundary condition requirements, the model was run at 2 cm stage increments for upstream water levels ranging from 9.6 m to 10.8 m (these represent the full range of flow conditions experienced within the reach). This procedure yielded mean annual deposition rates and particle-size distributions for each floodplain node.

Estimates of mean annual deposition rates over the period 1963–1992 were determined at four sites using measurements of the caesium-137 inventories of floodplain sediment cores. Caesium-137 is an artificial fallout radionuclide associated with the atmospheric testing of nuclear weapons during the period from the mid 1950s to the mid 1960s. Following deposition as fallout it was strongly adsorbed by clay particles in the surface horizons of the soil (Frissel and Pennders, 1983), and its subsequent redistribution in the landscape, in association with sediment particles, may be used to document rates and patterns of sediment movement (cf. Walling and Bradley, 1990). Figure 7.9 shows caesium-137 profiles associated with four floodplain sites A2 to D2. These sites correspond to point bar, abandoned channel, depression and floodplain locations respectively, and are different from those used in the measurement and prediction of deposition patterns at the event scale. Profiles were derived by slicing soil cores into a number of samples at approximately 2 cm intervals (depths have subsequently been normalized to compensate for variations in sediment density throughout the profile). After air drying and disaggregation, the caesium-137 content of the samples was measured by gamma spectrometry using a high-purity germanium detector equipped with a multi-channel analyser. Counting times were typically about 25 000 s and provided an analytical precision of the order of ±10% (2 s.d.). Mean rates of sedimentation over the period 1963–1992 were estimated by dividing the depth to the profile peak, which is assumed to represent the amount of deposition since 1963 (less a small distance corresponding to the downward migration of caesium), by the time elapsed since the peak rate of atmospheric fallout in 1963 (29 years).

Table 7.2 gives the estimated and predicted mean annual deposition rates for the four floodplain sites. Once again rates of deposition appear to be linked to distance from the main channel and susceptibility to ponding. The very high rate of deposition in the abandoned channel (site B2) is consistent with the findings of other workers (Lewis and Lewin, 1983) and results in part from the high frequency of inundation of this low-lying area. Figure 7.10 shows the measured and predicted *ultimate* grain-size distributions and the predicted *effective* grain-size distributions for the same four deposits. Again the difference between the *ultimate* distributions for the surface material of the core deposits is less noticeable than that between the *effective* distributions, with the most significant difference in the latter being the proportion of sand-sized material present. Deposit A2 contains a much greater amount of such material than the other three samples. This appears to be one area where the model has difficulty in making accurate predictions, probably because

TABLE 7.2 Estimated and predicted mean annual deposition rates

Deposit	A2	B2	C2	D2
Profile location	Point Bar	Abandoned Channel	Depression	Floodplain
Distance from channel	10 m	5 m	10 m	110 m
Estimated deposition rate	3.68 mm year^{-1}	10.35 mm year^{-1}	4.95 mm year^{-1}	0.00 mm year^{-1}
Predicted deposition rate	3.35 mm year^{-1}	14.53 mm year^{-1}	3.96 mm year^{-1}	0.05 mm year^{-1}

sand-sized particles make up only a tiny fraction of the river's sediment load, thus little information about this size class is available for model parameterization purposes. The coarsest *effective* sample is also clearly deposit A2, which corresponds to the point bar location and is composed of almost 70% > 63 µm particles. Deposits B2 and C2 represent samples of intermediate coarseness and both have similar *effective* size distributions, probably as a result of similarities in their proximity to the channel and susceptibility to ponding. Deposit D2 is significantly finer than all the other samples and exhibits the effects of exhaustion of >63 µm particles in the floodplain flow at great distances from the channel. Removal of the majority of sediment in the coarsest class from the water, prior to reaching site D, means that the 32–63 µm class has become the dominant source of sediment resulting in a finer *effective* size distribution and lower deposition rates.

CONCLUSIONS

Preliminary results from a two-dimensional finite difference model of flood hydraulics and floodplain sedimentation suggest that patterns of flow direction and inundation extent in complex topographic environments may be determined using a relatively simple hydraulic scheme. Furthermore, comparison of model predictions with measurements of deposition rates and particle-size characteristics of deposited sediment, over a range of temporal scales, indicates a reasonable degree of correspondence between measured and predicted rates and patterns. Results suggest that peak mean sedimentation rates may be of the order of 1–1.5 cm year^{-1} for abandoned channel fill sites, while more common rates range from 3–5 mm year^{-1} near the main channel to as little as 0.1 mm year^{-1} in floodplain areas distant from the river. Evidence of changes in the particle-size characteristics of flood deposits with increasing distance from the main channel suggest that, although both *effective* and *ultimate* grain-size distributions fine across the floodplain, this trend is significantly more marked for the former particles. These findings have important implications for the role of sediment aggregration in transport and deposition processes. They also call into question the validity of using size data based purely upon discrete primary particle distributions in studies of floodplain sedimentation.

ACKNOWLEDGEMENTS

The authors gratefully acknowledge the support of the UK Natural Environment Research Council in providing a Postgraduate Studentship (APN) for work on overbank sedimentation in Devon rivers. The cooperation of local landowners in permitting access to floodplain sites and the assistance of the National Rivers Authority in providing flow data for the Culm are also acknowledged with gratitude.

REFERENCES

Allen, J. R. L. 1965. A review of the origin and characteristics of recent alluvial sediments. *Sedimentology*, **5**, 89–191.
Bates, P. D., Baird, L., Anderson, M. G., Walling, D. E. and Simm, D. 1992. Modelling floodplain flows using a two-dimensional finite element model. *Earth Surface Processes and Landforms*, **217**, 575–588.
Brown, A. G. and Keough, M. 1992. Palaeochannels, palaeoland-surfaces and three-dimensional reconstruction of floodplain environmental change. In: Carling, P. A. and Petts, G. E. (eds), *Lowland Floodplain Rivers: Geomorphological Perspectives*, Wiley, Chichester, pp. 185–202.
Chow, V. T. 1959. *Open-Channel Hydraulics*. McGraw-Hill, New York, 680 pp.
Ervine, D. A. and Ellis, J. 1987. Experimental and Computational aspects of overbank flood-plain flow. *Transactions of the Royal Society of Edinburgh: Earth Sciences*, **78**, 315–325.
Fischer, H. B., List, E. J., Koh, R. C. Y., Imberger, J. and Brooks, N. H. 1979. *Mixing in Inland Coastal Waters*. Academic Press, London, 483 pp.
Frissel, M. J. and Pennders, R. 1983. Models for the accumulation and migration of ^{90}Sr, ^{137}Cs, ^{239,240}Pu and ^{214}Am in the upper layer of soils. In: *Ecological Aspects of Radionuclide Release*, Spec. Publ. Brit. Ecol. Soc. no. 3, pp. 63–72.
Gee, D. M., Anderson, M. G. and Baird, L. 1990. Large-scale floodplain modelling. *Earth Surface Processes and Landforms*, **15**, 513–523.
Howard, A. D. 1992. Modelling channel migration and floodplain sedimentation in meandering streams. In: Carling, P. A. and Petts, G. E. (eds), *Lowland Floodplain Rivers: Geomorphological Perspectives*. Wiley, Chichester, pp. 1–41.
James, C. S. 1985. Sediment transfer to overbank sections. *Journal of Hydraulic Research*, **23**, 435–452.
Kiely, G. 1990. Overbank flow in meandering compound channels: The important mechanisms. In: White, W. R. (ed.), *International Conference on River Flood Hydraulics*, Wiley, Chichester, pp. 207–217.
Lambert, C. P. and Walling, D. E. 1987. Floodplain sedimentation: A preliminary investigation of contemporary deposition within the lower reaches of the River Culm, Devon, UK. *Geografiska Annaler*, **69A**, 393–404.
Lewin, J. 1978. Floodplain geomorphology. *Progress in Physical Geography*, **2**, 408–437.
Lewis, G. W. and Lewin, J. 1983. Alluvial cutoffs in Wales and the Borderlands. *Special Publication of the International Association of Sedimentologists*, **6**, 145–154.
Marriott, S. 1992. Textural analysis and modelling of a flood deposit: River Severn, UK. *Earth Surface Processes and Landforms*, **17**, 687–697.
Parker, G. 1978. Self-formed straight rivers with equilibrium banks and mobile bed: I–The sand-silt river. *Journal of Fluid Mechanics*, **89**, 109–125.
Pizzuto, J. E. 1987. Sediment diffusion during overbank flows. *Sedimentology*, **34**, 301–317.
Samuels, P. G. 1985. Modelling of river and floodplain flow using the finite element method. Hydraulics Research, Tech. Report No. SR61, Wallingford, UK.
Shiono, K. and Knight, D. W. 1991. Turbulent open-channel flows with variable depth across the channel. *Journal of Fluid Mechanics*, **222**, 617–646.
Shotton, E. W. 1978. Archaeological inference from the study of alluvium in the Lower Severn–Avon Valleys. In: Limbrey, S. and Evans, J. G. (eds), *Man's Effect on the Landscape: The Lowland Zone. Council for British Archaeology Research Report*, **21**, 27–32.
Trimble, S. W. 1981. Changes in sediment storage in the Coon Creek Basin, Driftless Area, Wisconsin, 1853–1975. *Science*, **214**, 181–183.

Urban, C. and Zielke, W. 1985. Steady-state solution for two-dimensional flows in rivers with floodplains. *The Hydraulics of Floods and Flood Control*, Proceedings of the 2nd International Conference on Floods and Flood Control. Cambridge, pp. 389–398.

Walling, D. E. and Bradley, S. B. 1990. Some applications of caesium-137 measurements in the study of fluvial erosion, transport and deposition. In: *Erosion, Transport and Deposition Processes, Proceedings of the Jerusalem Workshop*, International Association of Hydrological Sciences Publication number 189, pp. 179–203.

Walling, D. E. and Woodward, J. C. 1993. Use of a field based water elutriation system for monitoring the *in situ* particle size characteristics of fluvial suspended sediment. *Water Research*, 27, 1413–1421.

Walling, D. E., Bradley, S. B. and Lambert, C. P. 1986. Conveyance losses of suspended sediment within a floodplain system. In: Hadley, R. F. (ed.), *Drainage Basin Sediment Delivery, Proceedings of the Albuquerque Symposium*, International Association of Hydrological Sciences Publication number 159, pp. 119–131.

Wright, R. R. and Carstens, M. R. 1970. Linear-momentum flux to overbank sections. *Journal of the hydraulics Division, ASCE*, **96**(HY9), 1781–1793.

Yen, B. C. and Yen, C. L. 1983. Flood flow over meandering channels. In: Elliot, C. M. (ed.) *River meandering Proceedings of the Conference Rivers '83*, pp. 554–561.

Yen, C. and Overton, D. E. 1973. Shape effects on resistance in flood-plain channels. *Journal of the Hydraulics Division, ASCE*, **99**(HY1), 219–238.

8

Hydraulic Geometry and Channel Scour, Fraser River, British Columbia, Canada

E. J. HICKIN

Department of Geography and the Institute for Quaternary Research, Simon Fraser University, Burnaby, BC, Canada

ABSTRACT

A simple model of at-a-station hydraulic geometry for a rectangular channel which scours above a threshold discharge is introduced to illustrate potential effects of scouring on the hydraulic geometry relations in rivers. Water Survey of Canada archival data are used to identify the threshold scour discharge (the transition from rigid to fully alluvial boundary conditions) for Fraser River at Marguerite, British Columbia. Data for channel width (w), water-surface elevation (E), flow depth (d) and mean velocity (v), are presented as conventional power-function relations and as untransformed linear relations, all dependent on discharge (Q). The latter reveal threshold scour-discharge-dependent discontinuities in E/Q, d/Q, v/Q and $v/d/Q$, consistent with model predictions but completely obscured by the corresponding conventional power-function relations. Thus, conventional data transformation is not recommended in studies of within-channel processes in rivers. Given that the threshold scour discharge appears to be identifiable from the hydraulic geometry of some river channels, several lines of further investigation are suggested.

INTRODUCTION

Principles of hydraulic geometry (Leopold and Maddock, 1953) recognize that, as discharge changes in a river, width, depth and velocity in the channel cross-section mutually adjust to accommodate the altered flow. These adjustments can be viewed as accommodations in time (at-a-station) or in space (downstream). "Downstream" hydraulic geometry refers these mutual adjustments to some formative discharge (usually bankfull flow) as it increases along a river as drainage area increases; it is interpreted as the dynamic and equilibrated response of a fully alluvial channel to the balance of flow forces and the resistance of boundary materials. "At-a-station" hydraulic geometry, on the other hand, refers these mutual adjustments in channel form to the changing stage at a given cross-section; it is interpreted as the largely

---

River Geomorphology. Edited by Edward J. Hickin
© 1995 John Wiley & Sons Ltd

passive accommodation of flows in an essentially rigid (non-alluvial) channel formed at some higher previous discharge (bankfull flow). Conventionally, hydraulic geometry relations are depicted as simple power functions of discharge although several researchers have noted that this convenient depiction is only approximate and may obscure physically meaningful departures from log-linearity (for example, see Knighton, 1975; Church, 1980). Nevertheless, the expedience of power-functions has focused thinking on the continuity of adjustments and has strongly influenced the theory of hydraulic geometry of river channels.

The purpose of this paper is to explore the nature of at-a-station hydraulic geometry in relation to a fundamental discontinuity in alluvial channels: the threshold transition from a fixed to mobile boundary. Physical reasoning dictates that at-a-station hydraulic geometry on an alluvial river must have at least two distinctly contrasting discharge domains. The first domain, ranging from zero discharge to that corresponding to the beginning of channel scour (threshold discharge), describes the hydraulic geometry of a fixed boundary. The second domain, ranging from threshold discharge to bankfull flow, describes a channel which is at least partly mobile.

Figure 8.1 illustrates schematically elements of a hydraulic geometry that one might expect in a rectangular channel with a well-defined threshold discharge (Q_t). As discharge increases to the point of scour at Q_t flow velocity and flow depth increase over a fixed bed as relative roughness declines. Because the bed is fixed, water-surface elevation rises at the same rate as the depth of flow increases. Once discharge exceeds the threshold (Q_t) for bed scour, however, the elevation of the bed is progressively lowered (unless armouring develops or an obstruction such as bedrock is encountered). Since scour cannot instantaneously accommodate rapidly

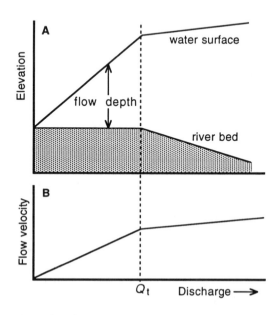

FIGURE 8.1 Schematic hydraulic geometry for a rectangular channel in relation to the threshold scouring discharge (Q_t)

increasing storm discharge and thus fully absorb the change in flow, flow depth is now accommodated by both water-surface elevation increases and lowering of the bed. Thus the rate of increase in water-surface elevation with discharge declines, producing a single discontinuity in dQ/dE at Q_t. In this example, the rate of change in flow depth also declines beyond Q_t along with flow velocity but the general case clearly depends on the relative rates of change in bed and water-surface elevations as discharge increases.

THE FRASER RIVER: THE FIELD SITE AND DATA

Ideal data to test for the presence of scour-induced discontinuities in hydraulic geometry are not as readily available as one might first suppose. A primary site requirement is that the channel boundary must begin to scour well before bankfull stage. Sand-bed rivers are likely candidates in this regard but they also are complicated by the effects of significantly changing bedforms. Gravel-bed rivers, on the other hand, often have regular planar beds but they also are often not mobile until discharges approach bankfull stage. Furthermore, most small rivers are not well suited to the purpose at hand because their typically flashy hydrology means that discharges greater than Q_t often are of such short duration that scouring of the bed is a distinctly disequilibrium phenomenon.

The site selected for this test is the Water Survey of Canada (WSC) gauging station at Marguerite on Fraser River in British Columbia (Figure 8.2). About half of the 232 000 km^2 Fraser catchment drains to this point on the river through a single and somewhat incised channel. Here the bed of Fraser River consists of sandy gravel, is seasonally mobile in response to snowmelt-induced sustained high flows, and is routinely monitored three to six times annually by WSC technicians in order to keep current the rating curve for the channel. Furthermore, Marguerite is the only gauging station on Fraser River which is not operated from a bridge with the attendant complication of structure-specific bridge-pier scour and fill; flow velocity and depth are measured from an aerial cableway.

Figure 8.3 shows a low-flow view of the Fraser River channel at Marguerite and several typical measured cross-sections for various discharges in 1986. At the mean annual flood ($Q_{2.33} = 4654$ m^3 s^{-1}) the river is about 220 m wide, averages 7.1 m in depth and flows at a mean velocity of almost 3 m s^{-1}.

The principal data source for this study is the WSC technicians' field notes held in the WSC archives in North Vancouver, British Columbia. Eight years of records (1980–1987) for a period which included a wide range of discharges were selected for analysis, yielding 50 complete sets of channel and flow measurements. Velocity and flow depth were measured on the same line of section from a bank monument at verticals spaced at approximately 5% width or 10 m. In the few cases where the presence of ice caused this practice to be modified for near-bank measurements, cross-sections were adjusted by interpolation so that all measurements were based on the same reference verticals. A common gauge-height datum applies to the entire data set. Velocity was measured with a cable-suspended current meter (at 10 cm intervals in the vertical) and flow depths by line sounding. Total measurement error for depth and velocity averages for channel cross-sections probably does not exceed ±5%.

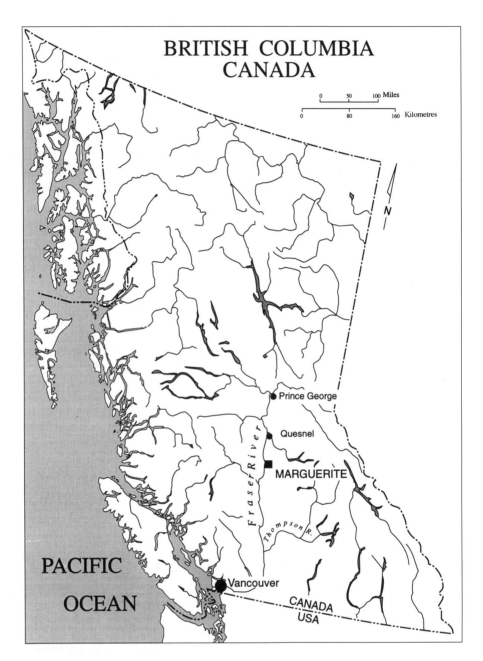

FIGURE 8.2 Location map of the Marguerite gauging station on Fraser River, British Columbia

Hydraulic Geometry and Channel Scour

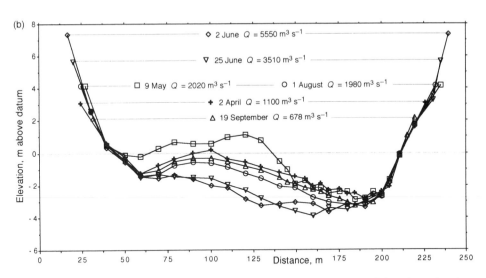

FIGURE 8.3 Fraser River at Marguerite gauging station. (A) A view of the channel at low flow. (B) Selected channel cross-sections for several discharges

Fieldwork at Marguerite was conducted to measure water-surface slope (by engineering transit) and to sample sediment from pits dug into exposed bars at low flow (February 1993).

Surface boundary materials at Marguerite station vary from year to year depending on the immediately preceding discharge history but generally consist of sandy gravel with particle size ranging from 0.06 mm to >64 mm but averaging 20–30 mm ($D_{50} = 23$ mm) (Carson, 1988). A limited and rather inadequate sample of bedload collected and analysed by WSC suggests that the bedload D_{50} falls in the same range as the bed material itself. Carson (1988) estimates that, at high flows, about 14% of the bed material is entrained in suspension, the remainder moving as bedload. Pits dug in the channel bars exposed at low flow reveal that their surfaces are moderately armoured as a result of waning-flow winnowing of fines.

DATA ANALYSIS

Hydraulic geometry

The at-a-station hydraulic geometry for Fraser River at Marguerite is shown in the usual way in Figure 8.4. Relations between width, depth and velocity and the independent variable discharge, are well described by simple power-functions ($R^2 > 0.94$) and might be taken as yet another confirmation of the assumptions of conventional hydraulic geometry. Fraser River is rather canal-like here with little

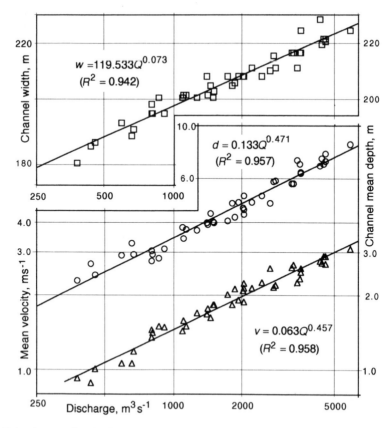

FIGURE 8.4 At-a-station hydraulic geometry for Fraser River at Marguerite (1980–1987), all scales are logarithmic

variation in width and most (~95%) of the increase in discharge being more or less equally accommodated by changes in flow velocity and depth. It will be argued below, however, that the continuous logarithmic relations are more apparent than real and that the log-transformation of the data obscures important processes and their effects on the channel.

Bed deformation

Mean bed elevation is calculated as $GH - d$ where GH is the gauge height (or water-surface elevation with respect to the gauging-station datum) and d is the average depth of flow over the bed (excluding bank zone). The measured record of mean bed elevation and mean daily discharge (Figure 8.5) for the eight-year test period is markedly periodic on the seasonal time-scale and out of phase. Annual peak (mean daily) freshet discharges in May–July varied from about 3000 m^3 s^{-1} to a maximum of 5570 m^3 s^{-1} (on 3 July 1986) while winter flows typically were less than 500 m^3 s^{-1}. As discharge increases to the seasonal peak flow during the freshet the river bed is scoured to a seasonal low while during declining autumn flows the bed fills and reaches its highest elevation during late winter. In high flow years (1985 and 1986, for example), the bed may be lowered from previous winter levels by as much as

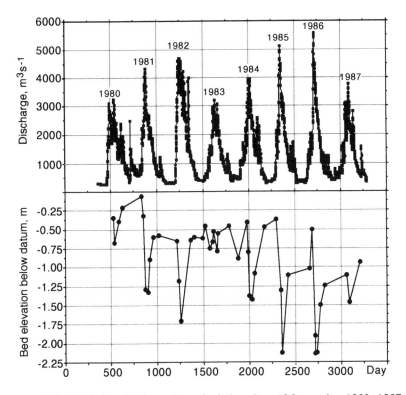

FIGURE 8.5 Variations in Fraser River bed elevation at Marguerite, 1980–1987

1.75 m on average while deformation is much more modest for smaller freshets (about 30 cm in 1983, for example).

The successive cross-sections of the Fraser River channel at Marguerite shown in Figure 8.3b exhibit the typical pattern of boundary adjustment to the seasonal discharge hydrograph, in this case for the high-flood year 1986. Discharge on 2 April, recorded after the seasonally increasing flow had cleared ice from the station, was 1100 m^3 s^{-1} and the average bed elevation stood at -1.03 m above datum. As discharge increased to 2020 m^3 s^{-1} on 9 May, the bed filled and the mean bed elevation increased to -0.505 m above datum as sand was moved from deeper parts of the channel into storage in large longitudinal sand sheets. A further increase in discharge to 3510 m^3 s^{-1} on 25 June, however, initiated general scour and the bar sediments were removed along with coarser basal sediments, lowering the average bed elevation to -2.12 m above datum; peak local scour during this period amounted to 3.4 m (almost 2 m of basal gravels below the surficial sand sheet). Discharge continued to increase until 2 June (5550 m^3 s^{-1}) although average bed elevation changed little with minor zones of cut in the area of the former bar balancing fill in the thalweg.

By 1 August discharge had declined to 1980 m^3 s^{-1} and the bed had filled again to stand at -1.5 m average bed elevation, about 0.5 m above the lowest recorded average level for the year; locally the fill reached almost 2 m in this phase. A further decline in discharge to 678 m^3 s^{-1} on 19 September was accompanied by continued filling of the bed to an average elevation of -1.24 m above datum, virtually back to its earlier 2 April position.

Measurements of winter bed elevations (and discharge) are complicated by the presence of ice on the river but it is clear from observation that sand accumulates in the channel during this period, causing some additional increase in mean bed elevation until the cycle of cut and fill is repeated in the following year.

Threshold scour discharge

Although the pattern of discharge and bed elevation depicted in Figure 8.5 clearly reveals seasonal alternations of bed scouring and filling, the precise discharge or gauge height at which scouring begins each year is not obvious because the bed-elevation series is not continuous.

Figure 8.6A shows a simple plot of bed elevation versus discharge but the relationship is poorly defined. There are several reasons for the scatter. First, the data include both scouring and filling events and hysteresis between discharge and bed elevation appears to be common. Second, the bed elevation measured on the day of the survey may well be partly a function of a preceding higher discharge. A further complication is that scouring may not start from precisely the same winter bed elevation in all seasons although the available measurements cannot resolve this question. Nevertheless, Figure 8.6A does suggest that bed elevations tend to decline at discharges higher than about 2500 m^3 s^{-1}.

If the data are generalized as a five-point moving average of bed elevation, the onset of scour is more clearly evident (Figure 8.6B). Once again, bed elevation clearly falls rapidly when discharge exceeds about 2500 m^3 s^{-1}.

Hydraulic Geometry and Channel Scour 163

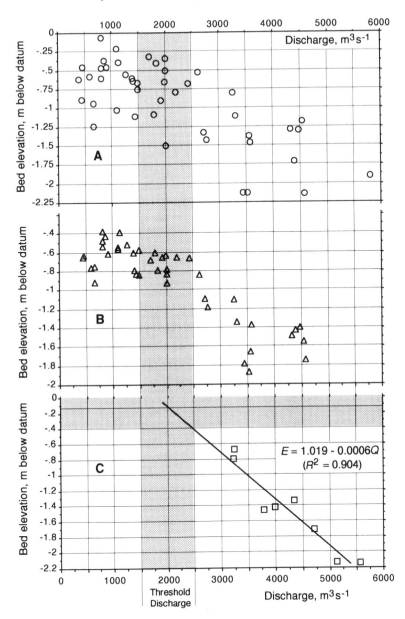

FIGURE 8.6 Bed elevation versus discharge for Fraser River at Marguerite

If the minimum bed elevation reached by the seasonal scouring event (characterized by a continuous decline in bed elevation on the rising limb of the seasonal hydrograph) is correlated with the peak seasonal discharge, a strong inverse linear trend is evident (Figure 8.6C). This trend can be extrapolated to the winter bed level to provide a further indication of the threshold scouring discharge. Because few winter measurements have been obtained it is not entirely clear where the average winter bed elevation occurs but the data suggest that it is within a few decametres (0–40 cm) of the gauge datum. These relations imply a threshold discharge for scour of about 2000 $m^3 s^{-1}$.

Hydraulic relations and threshold scour discharge

The relations of various hydraulic and channel parameters to discharge for the Fraser River at Marguerite are shown as simple bivariate plots in Figure 8.7; the curves shown have been fitted by eye to highlight the behaviour of the river in the two flow domains (rigid boundary versus alluvial boundary).

Clearly, there are discontinuities evident in these plots, all but one of which correspond with the threshold scour discharge of about 2000 $m^3 s^{-1}$. The exception is the plot of channel width versus discharge, which although displaying a discontinuity, is "kinked" at a much lower discharge (~1000 $m^3 s^{-1}$) and maintains a linear trend across the threshold scour discharge. This observation is important to the discussion to follow because once the inset low-flow channel of the Fraser is filled (the cause of the width/discharge "kink" at about 1000 $m^3 s^{-1}$), further increases in discharge are associated with a continuous change in channel width, consistent with the canal-like trapezoidal channel section at this station on Fraser River. That is, there is no discontinuity in channel width encountered at 2000 $m^3 s^{-1}$ which could explain the discontinuities evident in the other trends shown in Figure 8.7.

The rating curve for the station (Figure 8.7A) can be described by two linear trends consistent with those postulated in Figure 8.1A. As discharge increases and fills the essentially rigid boundary of the channel, the water surface rises at a rate reflecting changes in the wetted channel shape and resistance to flow. At about the 2000 $m^3 s^{-1}$ threshold discharge, further increases in discharge are associated with less rapid increases in water-surface elevation because, as discussed earlier, scour begins to lower the bed and the increases in flow depth dictated by continuity are now a function of changing bed elevation as well as changes in water-surface elevation.

Flow depth versus discharge (Figure 8.7C) displays a similar but far less pronounced change in gradient across the threshold discharge. This modest change reflects the accompanying and compensating changes in the gradients of water-surface elevation and bed elevation (see Figure 8.6) which have both shifted in the same direction as scouring increases.

The marked decline in the mean velocity/discharge gradient at the threshold scour discharge is quite clearly evident in Figure 8.7B, once again consistent in trend with that postulated earlier (see Figure 8.1B). The marked decline in the rate of increase in mean velocity with discharge when the threshold scour discharge is exceeded, suggests that increases in resistance to flow may be associated with scouring of the bed (unfortunately water-surface slope measurements over the relevant range of discharge are not available). Such a circumstance might be expected since scouring is spatially irregular and, in addition, scouring very likely leads to roughening of the boundary by exhuming large basal bed material.

Hydraulic Geometry and Channel Scour

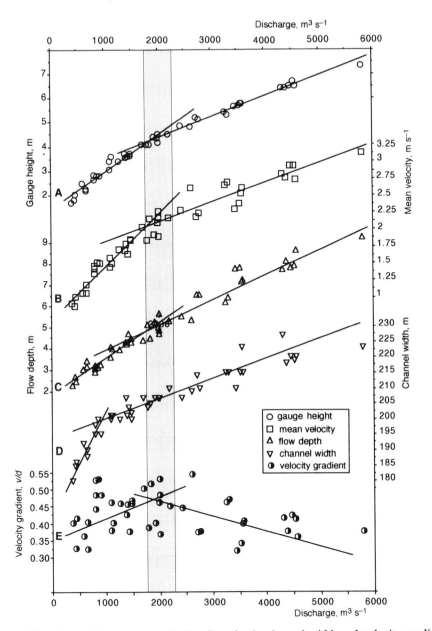

FIGURE 8.7 Gauge height, mean velocity, flow depth; channel width and velocity gradient, in relation to threshold scour discharge (~2000 m³ s⁻¹) for Fraser River at Marguerite

The rapid decline in the velocity/discharge gradient (v/Q) across the threshold scour gradient, accompanied by a much more modest decline in the flow depth/discharge gradient (d/Q), has a net effect of reversing the discharge-dependent trend in the mean velocity gradient (v/d) from positive to negative across the threshold scour discharge (Figure 8.7E).

CONCLUSIONS

These data indicate that the process discontinuity represented by the scour threshold on Fraser River at Marguerite is reflected in the hydraulic geometry of the channel. There is no reason to suppose that similar discontinuities are not a feature of all scouring (alluvial) channels although it will be most evident in larger rivers where discharge varies regularly on a seasonal basis, circumstances which ensure full equilibration of the channel with the controlling discharge. Certainly, elsewhere on Fraser River, a well-defined threshold scour discharge also appears to be evident in the seasonal changes in mean bed elevation (Sichingabula, 1993).

The scour-based discontinuities identified here are completely obscured by the conventional presentation of hydraulic geometry in which the data are logarithmically transformed (Figure 8.4). This observation suggests that, although power-function hydraulic geometry may be useful for some purposes, the data transformations on which it is based probably should be avoided if the within-channel processes are the subject of investigation.

Many interesting questions about hydraulics, sediment supply and transport in relation to scour thresholds in rivers are raised by this pilot study and invite further work, including among others:

(i) Is there a discontinuity in resistance to flow at the threshold scour discharge?
(ii) What is the relation of bankfull discharge to threshold scour discharge?
(iii) What is the relation of the threshold scour discharge to that causing incipient motion of bed material?
(iv) Since channel scouring releases locally stored sediment in the channel, is there a similar measurable discontinuity in suspended-sediment rating curves?
(v) Is there sedimentological evidence of this seasonal pattern of cut and fill recorded in the basal sediments of floodplains of rivers characterized by this behaviour?

The answers to these and other related questions are of considerable scientific interest but some of them may be of even greater engineering significance. For example, the prospect of predicting the flow conditions corresponding to incipient bed-material motion from hydraulic geometry relations is one which clearly demands further attention. This paper will have served its purpose if others are persuaded to consider these matters further.

ACKNOWLEDGEMENTS

This study is part of a research programme funded by Simon Fraser University and the Natural Science and Engineering Research Council of Canada.

REFERENCES

Carson, M. 1988. Sediment station analysis: Fraser River near Marguerite. Inland Waters Directorate, Environment Canada, 280 pp.

Church, M. 1980. On the equations of hydraulic geometry: Department of Geography, The University of British Columbia, Vancouver, British Columbia, 93 pp.

Knighton, A. D. 1975. Variations in at-a-station hydraulic geometry: *American Journal of Science*, **275**, 186–218.

Leopold, L. B. and Maddock, T. 1953. The hydraulic geometry of stream channels and some physiographic implications. *United States Geological Survey, Professional Paper*, **252**, 56 pp.

Sichingabula, H. M. 1993. Character and controls of suspended-sediment concentration and discharge effectiveness, Fraser River, British Columbia, Canada. PhD thesis, Simon Fraser University, 358 pp.

9

Torrential Flow Frequency and Morphological Adjustments of Ephemeral Channels in South-East Spain

C. CONESA GARCÍA

Department of Geography, University of Murcia, Spain

ABSTRACT

This work focuses on the morphological adjustments of ephemeral channels or those of very irregular regimes in south-east Spain, to the frequency of the effective events in terms of total sediment transport, bankfull discharges and floods. The chosen reaches belong to fluvial systems in semi-arid environments and represent channel patterns ranging from low sinuosity to braided and meandering. The balance between erosion and sedimentation in most channels has been changed greatly by the high degree of regularization and protection against floods in the Segura River, reducing as well its torrential regime. Therefore, it has been necessary to choose marginal reaches not affected by the regularized flows and streams more or less separated from the main diversion works, reservoirs and dams.

The identification of geomorphic processes and thresholds was carried out taking into account two basic types of events: major events, causing overall changes in the fluvial system, and moderate events, which only cause local variations within the forms produced by the major ones.

The field investigation (1985-1991) coincided with years of intense drought, and less frequent floods than in previous decades. Nevertheless, their magnitude was enough to cause, within dynamic equilibrium, important morphological adjustments, especially in high-risk flood areas.

INTRODUCTION

In the last few decades, the adjustment of fluvial systems to environmental change has been of particular concern to researchers in fluvial geomorphology. The 10th meeting in the series of annual geomorphology symposia sponsored by the Department of Geological Sciences and Environmental Studies, State University of New York at Binghamton (1979), was devoted to this topic. The proceedings, edited by Rhodes and Williams in 1982, included important contributions to advancing the understanding of river systems. Among these is the work of Harvey *et al.*, which refers to event frequency as it influences process thresholds in upland Britain. Their plan of study, analysing first of all the geomorphic responses of the channel to different hydrologic thresholds, and then the occurrence of each type of event, leading to a discussion on its dynamic equilibrium, is adopted in this paper, but the results obtained are very different. For this reason it would be as well to ask what the effects of the magnitude and frequency of torrential events are, how far-reaching are the effects of the overall adjustments caused by major events, and to what extent the ephemeral channels maintain a dynamic equilibrium in this semi-arid context. In these conditions, new thresholds and event types, different from those established in more humid environments (Dury, 1973; Wolman and Gerson, 1978; Harvey *et al.*, 1982), may emerge.

Other contributions to the understanding of dynamic adjustments of fluvial channels which are particularly significant are based on the unit stream-power concept and the theory of minimum rate of energy dissipation (Yang, 1983; Yang and Song, 1986; Yang and Molinas, 1988). The former concept was used to derive the functional relationship between total sediment concentration and unit stream power. The theory of minimum rate of energy dissipation states that an alluvial channel is in equilibrium if its rate of energy dissipation is at its minimum under the given climate, hydrologic, geologic, and man-made constraints. "The minimum value depends on the constraints applied to the system" (Yang and Song, 1982).

The applicability of these conceptual frameworks to fluvial dynamics in semi-arid streams takes on undoubtable interest, in spite of the difficulty of re-adapting its equations to the available field data. In this type of channel, where adjustments to individual events are so extreme, and recovery periods so long, it is equally useful to know the different response to floods as a result of different discharge ranges and event frequency. Work has already been carried out along these lines by Pickup and Rieger (1979), according to whom every competent event has some influence on channel form, so that channel form at a given time is a weighted sum of all input discharge effects up to and including that time.

Up to now, the study of ephemeral channels in arid and semi-arid areas was broached from the point of view of their own morphological and sedimentary dynamics, with little general classification of events. If we consider episodic fluvial sedimentation in channels of this type, McKee *et al.*'s studies (1967) on flood deposits in Bijou Creek (Colorado), Glennie's (1970) on desert environments, Tunbridge's (1984) on ephemeral streams of north Devon, and Castelltort and Marzo's (1986) on the "Catalánides" (north-east Spain) stand out, among others.

In extremely arid zones it is worthwhile highlighting the paper by Nouh (1988) on regime channels. His field observations were collected from 37 ephemeral channels in Saudi Arabia, but the data obtained were used to develop regime formulae relating the channel geometry to the characteristics of flash flood and sediment flow, and not to establish a typology of morphological changes as a function of hydrologic events. In the same line of research is the study by Clark and Davies (1988), devoted to the application of regime theory to wadi channels in arid conditions of PDR Yemen.

This paper focuses on the effectiveness of torrential streams south-east Spain having widely different magnitude–frequency relationships.

STUDY AREAS

An important environmental problem in south-east Spain is the high erosion rate (40 to 120 t ha^{-1} year^{-1}), a function of the semi-arid conditions (high temperature, scant and torrential rains), the steep slopes of the Betic Range, the large areas of erodible lithologies (Neogene–Quaternary marls and clays), the sparsely vegetated terrain, and human impacts on the landscape.

A whole group of fluvial systems (the Jucar, Segura, Almanzora rivers), consisting of several courses of permanent water and dense networks of ephemeral channels ("*ramblas*" and gullies) is responsible for the liberation of vast quantities of sediments from the substantial soil loss which occurs in these basins. Channels, in this region, therefore, are settings for extreme morphological dynamics. Flow regime is very irregular and torrential. Channels which are dry for more than 6 or 7 months may suddenly carry, in just a few hours, more than 200 m^3 s^{-1}, and even above 1000 m^3 s^{-1} in major floods. The flow then reaches a substantial velocity and transports a high suspended load.

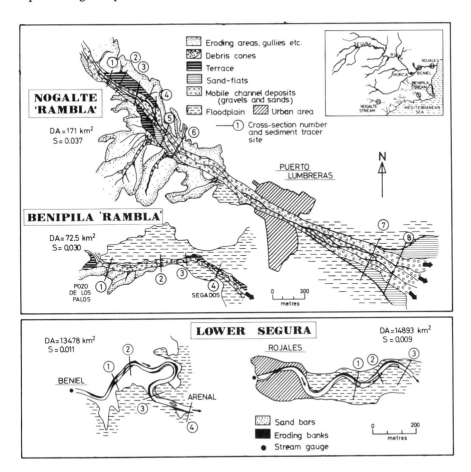

FIGURE 9.1 Location of the study sites in the regional context

FIGURE 9.2 The study sites. (a) The Nogalte *Rambla* downstream of Puerto Lumbreras, (b) the Benipila *Rambla* at Pozo de los Palos, (c) the Lower Segura downstream of Beniel

In this paper, the morphological adjustments caused by these torrential flows will be analysed. For this purpose we have chosen several sections of the ephemeral channel ("*rambla*"), within the Region of Murcia, and some sites in the Lower Segura River (Figure 9.1 and 9.2). The latter has experienced a significant decrease in discharge because of intense use before the river flows through the city of Murcia and because of the construction of the "Regueron", a diversion channel aimed at irrigation and flood protection. In all the sections monitored, we have examined the relationships between events of differing magnitudes and frequencies in the context of adjustments within dynamic equilibrium.

The "*ramblas*" studied, Nogalte and Benipila, have experienced significant morphological changes in the stream channel and floodplain. The basin of the Nogalte *Rambla* is situated in the interior band of the Betic Range. Its area (171 km^2) consists of two well-differentiated sections: a broad headwater composed of conglomerates, phyllites and mica-schists, on which a dense drainage network has developed, and a lower reach, where the Quaternary fill of a great tectonic depression is found; this depression is limited at north and south by longitudinal faults of direction ENE. Before it reaches the bottom of the depression, the Nogalte stream has a braided pattern, in which two study sites were chosen. The Benipila *Rambla* drains a basin of 72.5 km^2, located on coastal Betic area. Upstream it cuts across metamorphic and lime-dolomitic terrain from the Permo-Triassic period, and downstream it flows through detritic post-orogenic sediments. The sections monitored in the *rambla* correspond to this last reach.

CARTOGRAPHIC BASE

The characterization of the selected areas, which correspond to the Lower Segura (Beniel–El Arenal) and to the Nogalte and Benipila *ramblas*, and the representation of the overall morphological changes produced during the period of observation for each of these fluvial systems, were carried out on the cartographic base provided by aerial photographs dating from 1978 (Instituto de Reforma y Desarrollo Agrario, Spain; scale 1/18 000) and the orthophotomaps and line plans (1988) (Cartographic Service of the Comunidad Autónoma de Murcia; scale 1/5000)(No. 975/1–5 and 2–6, No. 977/4–1 and 5–2).

In the case of the Segura River in Algorfa–Rojales, aerial photographs (1979), the property register map (1/5000) and the topographic map (PGOU) (1/10000) (1981), were used.

DATA AND FIELD MEASUREMENTS

Bed scour-depth in the ephemeral channels was obtained by the scour-chain technique (Leopold *et al.*, 1966), and bed-material movement was monitored by tracers, setting the thresholds that are specified in Tables 9.2 and 9.3. The tracers used were gravels and cobbles collected from the channels and painted in a distinctive yellow colour using fluorescent paint. Tracers were prepared in sets of 100 and placed in the channel along controlled cross-sections in the Nogalte and Benipila *ramblas*. Care was taken to ensure that each set of tracers represented the range of bed-material sizes and shapes. The sampling of the suspended material was carried out by a depth-integrated sampler for the Segura in Algorfa and Beniel and by a "level sampler" with fixed platform, designed from the Hayim models (Schick, 1967, Lekach and Schick, 1982), for the torrential flows in ephemeral channels.

Bed-material characteristics, bar location, planform shape, micro-topography and bar unit components were examined. The size distribution of material on bar surfaces was determined since this plays a significant role in affecting the entrainment processes.

Bank erosion and the morphology of the stream bed were monitored using erosion pins and by comparing photographs taken after each torrential event, with an observation period of seven years for the Benipila *Rambla* (1985–1991) and of six years for the sections of Nogalte and Lower Segura (1986–1991).

Floodpeak discharge has been obtained from the Chèzy–Manning formula using the maximum height reached by the flow. The sections at Beniel and Algorfa were gauged by the Confederación Hidrográfica del Segura.

FREQUENCY AND PROBABILITY OF THE TORRENTIAL EVENTS

Ephemeral-channel morphology depends on the discharge regime in the range of competent discharges. These may be grouped in the following categories:

Class 1. Overbank flows which affect the whole fluvial system. In the case of the *ramblas*, they greatly modify the floodplain, by producing at the same time both deposition and scour. Floodwater velocities over this surface are often spatially variable and may, in places, be high enough to produce scour rather than deposition. In Lower Segura they contribute to the vertical accretion of the floodplain.

Class 2. Dominant events controlling channel form. These comprise moderate discharges causing appreciable net changes in the bed and banks, and flows which just fill the section of alluvial channels without overtopping the banks (bankfull conditions).

Class 3. Discharges sufficient to cover the whole of the main channel and produce bedload movement.

Class 4. Very low energy flows, incapable of causing bedform adjustments.

In Table 9.1 the occurrence and probability of the events corresponding to these classes are shown. In the case of the Segura Beniel–El Arenal and Algorfa–Rojales sections, this probability was directly calculated from available gauge data.

Comparison of the discharge thresholds (q) with the maximum daily rainfall (p) values registered upstream in the catchment of the ephemeral channels, yields a high correlation coefficient between both parameters ($r = 0.89$), with adjustment defined by the equations $q = 0.596\ p + 19.36$ (Nogalte *Rambla*) and $q = 0.736\ p + 19.37$ (Benipila *Rambla*). The 20–30 year rainfall record has been used to estimate the mean annual flood-event probability.

TABLE 9.1 Event class threshold frequency

	Class	Discharge (m³ s⁻¹)	Occurrence No. of events	(times/ year)	Recurrence interval (years)	Annual exceedance probability (%)
Nogalte *Rambla*						
	1	106.0	1	0.17	6.00	0.27
	2	36.3	3	0.50	2.00	0.82
	3	7.0	9	1.50	0.67	2.46
	4	0.9	20	3.30	0.30	5.48
Benipila *Rambla*						
	1	85.0	2	0.29	3.45	0.55
	2	27.4	5	0.71	1.41	1.37
	3	5.2	11	1.57	0.64	3.01
	4	0.6	23	3.28	0.30	6.30
Segura River (Beniel – El Arenal)*						
	1	98.3	4	0.80	1.25	1.10
	2	30.7	15	3.00	0.33	4.11
	3	7.6	58	11.60	0.05	15.89
Segura River (Algorfa)**						
	1	166.0	2	0.40	2.50	0.55
	2	43.2	17	3.40	0.29	4.66
	3	9.8	55	11.00	0.09	15.07

* Discharge in Beniel
** Discharge in Rojales, 7 kms downstream of Algorfa

Except for the section of El Arenal, which represents a more vulnerable environment, the major thresholds (those implying an overall modification of the fluvial system class 1) are exceeded each 2 to 6 years (Table 9.1). Generally, they imply overflow and, in extreme situations, they bring about considerable modifications of the pre-existing channel pattern. In spite of the small series of stream gauges used in the fieldwork, the recurrence interval of this type of event, based on short-term gauging-station records, in the cases of Beniel and Rojales, is consistent regardless of the frequency distribution used (Gumbel (G), Log-Gumbel (LG) and Log-Pearson type III (LP))(Figure 9.3).

The thresholds that define morphological adjustment (class 2) represent torrential phenomena of a moderate intensity and duration, produced by one or more storms. These thresholds are established from instantaneous flood peaks of 30 to 43 m³ s⁻¹ and have a 1.4 year recurrence interval for the sections of *rambla* and a 3.5 to 4 month recurrence interval for the sections of the Segura studied (characterized by pronounced and lengthy low water stages).

The minor thresholds, frequently related to frontal rains of low intensity or isolated storms, bring about only slight modifications in the channel. Within this group, the events that cause some type of modification in the bedforms occur 2 to 12 times per year.

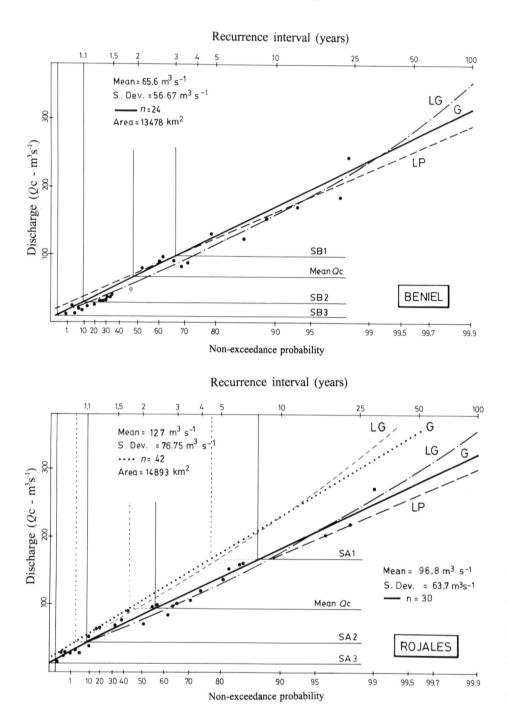

FIGURE 9.3 Recurrence intervals of the floodpeak discharge of the Segura River in Beniel and Rojales. Representation of thresholds for each event class

GEOMORPHOLOGICAL EFFECTIVENESS

The erosive processes caused by torrential runoff are especially significant in southeastern Spain, not only because of their enormous capacity for bed scour and bank erosion, but also because of the large amount of suspended load which they bring to the channels and the consequent modifications which this represents to the fluvial systems (Figure 9.4). High-water stages often exceed geomorphic thresholds. Moderate events have a limited effect on the morphology of the interior channel and only cause substantial modifications in particular circumstances.

Sediment transport

Figure 9.5 shows histograms of sediment load and discharge in the semi-arid streams studied. Magnitude–frequency analysis of these discharge events suggests that most sediment transport is achieved by intermediate to major runoff events, of low frequency. Flow regimes are extremely variable. The usually dry beds of the *ramblas* flash-flood when, on occasions (usually in autumn), high discharge is caused by exceptionally heavy rainfall. About 65% of sediment transport corresponds to extreme events, whose frequency in the period analysed was only about 5% of the total events. Sediment type also affects discharge effectiveness. Bedload transport requires the exceedance of a threshold stream power, so the most effective discharge should be more extreme (Richards, 1982). In the *rambla* sections investigated, the recurrence interval of the effective discharge ranged from 1.3 to 1.8 years, less than the bankfull discharge (whose recurrence interval ranged from 2 to 3.3 years).

FIGURE 9.4 Nogalte upstream from Puerto Lumbreras, carrying a large suspended-sediment load (9 September 1989)

The bedload measurements are considerably higher in the Nogalte braided stream (>2000 tonnes for discharge over 100 m^3 s^{-1}). This discharge threshold has only once been exceeded in the period 1986–1991, which means a slightly lower frequency than in previous decades. Suspended-sediment load increases progressively in accordance with the magnitude of the torrential events (Figure 9.5),

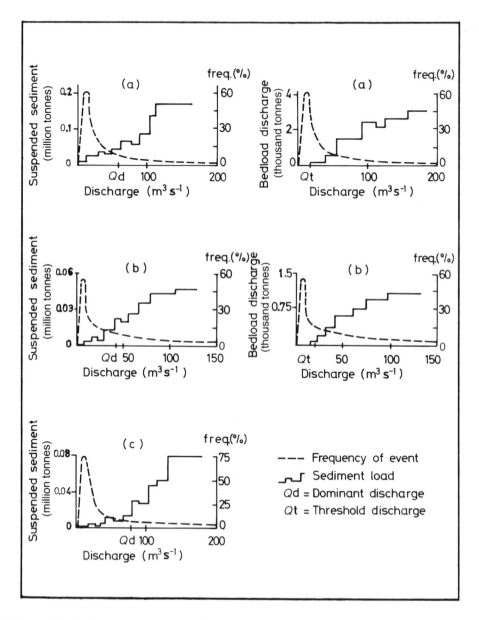

FIGURE 9.5 Relationship between sediment transport and event frequency for (a) Nogalte *Rambla*, Puerto Lumbreras; (b) Benipila *Rambla*, Los Segados; and (c) Lower Segura

and only on the Lower Segura tends to become stable once bankfull discharge has enreached. Generally, major events of low occurrence (one every three years on average) give a suspended load three to four times greater than that of moderate events. In the latter case, anomalies occur in which some discharges, while being smaller than others, may carry a greater suspended load. Such anomalies depend basically on the size and environmental characteristics of the area affected by the storm. However, a characteristic common to all the events is the non-coincidence of the peak of suspended load and flood discharge (hysteresis). This may be partially explained by the fact, already pointed out by Einstein and Shen (1964), that at the beginning of many storms the water moving towards and through the stream channels may find much more loose material ready to move than at the end of the storm.

Channel adjustments

The effects of sediment transport by torrential runoff are shown very clearly in the morphology of ephemeral channels and therefore should be systematically related to event magnitude and frequency. The class of extreme events determines channel capacity, affecting channel pattern, and more frequent events, although at times widely separated in time, control bedload movement and bedform adjustment.

Important adjustments of channel form and floodplain development are reflected in the bankfull capacity of the cross-section and the relationship of floodplain sedimentology and construction to channel pattern. The depositional activity of overbank flow on the floodplain is very different among the systems analysed. In the Nogalte braided stream the aggradation facies of the floodplain, downstream from Puerto Lumbreras, reflects lack of supply of fine sediments, in contrast to the floodplain of the Lower Segura, whose material consists almost entirely of silts and clays. In the course of the study, overbank flow caused the deposition of 5–25 cm of sand and silt in the sections close to the *rambla* sites (e.g. Nogalte sections 7 and 8). In those sections the deposition produced during flood hydrograph recession and in moderate events is greater than the erosion caused by high stream power in rising flows.

For the same period, more than 65% of the perimeter of the cross-sections monitored experienced an accretion of more than 5 cm, in some cases (sand flats and local bars) reaching as much as 25 and 30 cm. The segments of the perimeter which hardly showed minimal morphological changes make up 10–20%, while the erosion reaches are concentrated preferentially along the main channel banks, terrace edges and contact points of sand deposits with gravel bars (Figure 9.6).

On the Lower Segura erosion is of little importance. Here there is appreciable deposition of the two main types of sediments described by Allen (1982) for the meandering pattern: channel and overbank deposits. The frequent flooding between January 1986 and November 1991 caused considerable vertical accretion in the floodplain sectors nearer to the bed (>15 cm), greater than the lateral accretion of the point bars (5–15 cm).

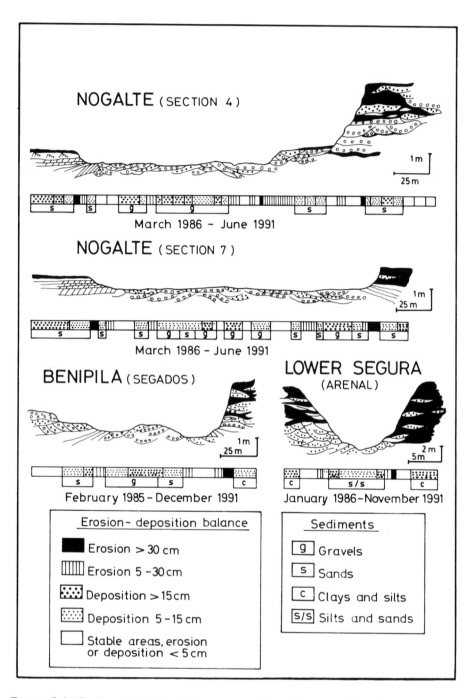

FIGURE 9.6 Erosion–deposition balance at some channel cross-sections of the study sites

CLASSIFICATION OF CHANNEL-FORMING EVENTS

In the Nogalte *Rambla*, the discharge thresholds established for the different classes of geomorphic events have been identified on the basis of the adjustments produced in channel sections of a braided pattern, upstream and downstream from Puerto Lumbreras. Table 9.2 shows the main types of channel adjustment caused by torrential events of different intensity. Discriminant analysis of the extensive information gathered allows the differentiation of four modalities of channel response in terms of the flow characteristics, during and immediately after the storm.

The changes in the channel associated with classes N1 and N2 and the threshold between both are presented tentatively, given the low frequency of extreme events in the period under analysis (1986–1991). On the other hand, this situation is normal in the sporadic and torrential watercourses in south-eastern Spain. These changes amount to just 12% of the total of events, but they control the channel morphology, since they represent the greater adjustments and removals of material.

The degree of braiding increases markedly above the threshold of class N1, also coinciding with the decrease (in low-water conditions) of the ratio between the space occupied by the emerged bar segments (b) and the accumulated width of the primary and secondary channels (ch) ($b/ch = 4.03 – 4.41$). In high regime conditions, sediment is transported as bedload and in suspension, the quantity of solid material removed being extremely high when bankfull discharge is exceeded.

Above the threshold of class N1, overflowing of the main channel and accretion of sand-flats are to be seen (Figure 9.7). These accretion deposits, left by episodic floods, display sets of planar and tabular cross-stratification and mixed forms at the surface, composed of bars, sand waves and ripples.

It is a braided system, formed by a group of very shallow and unstable channels that frequently change position, producing alternate aggradation and degradation related to the erosion of the lateral sand-flats (class N2).

The flows with a discharge ranging between 8 and 36 m^3 s^{-1} involve small local adjustments in the morphology of the channel (Class N3). They cause minor

TABLE 9.2 Channel adjustment classification in the Nogalte *Rambla*, 2 km upstream of Puerto Lumbreras

Class N1	Greatest channel modifications. Important variations in channel planform and cross-section. Accentuation of braided character. Very intense lateral erosion. Large suspended-sediment load. Destruction of whole bars of gravel and formation of new ones. Flooding and bank erosion of alluvial terraces.
Class N2	Variation of the braided structure without any substantial modification of the perimeter. Erosion of the sand flats and marginal deposits of the channel. Important texture changes of the bed bars. Coarse sediment transport (median size = 7–15 cm) and dense load in suspension.
Class N3	Slight channel modifications. Movement of gravels and cobbles (median size <7 cm). Modification of the gravel bar-tails and erosion of the lateral sand bars.
Class N4	Sand movement. No changes are observed in the channel morphology or in the texture of the bed deposits.

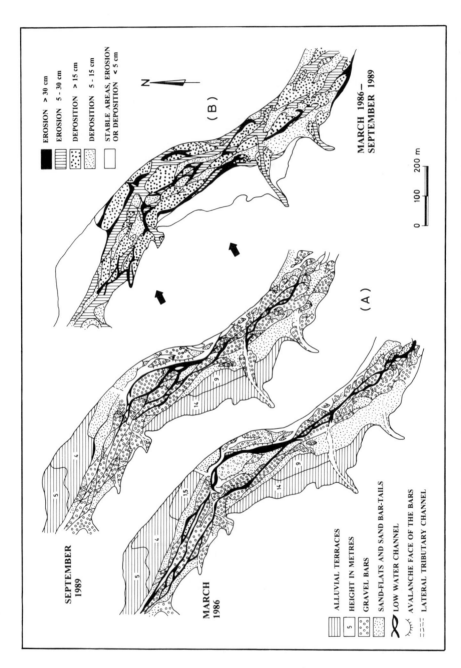

FIGURE 9.7 Nogalte *Rambla*. Morphological adjustments (A) and erosion–deposition balance (B), 1 km upstream of Puerto Lumbreras, March 1986–September 1989

transport of the bedload (gravels and cobbles) at flow velocities of 0.4–0.9 m s^{-1}, enough to develop straight-crested sand waves or megaripples and cause the lateral movement of gravel sheets with size segregation on the supraplatform of the simple bars.

The heads of these bars become areas for residual deposits of a differentiated texture, as is usual in highly unstable braided streams (Bluck, 1982). Small furrows begin on these surfaces which finally converge into a marginal secondary channel.

In Benipila *Rambla*, the initiation of erosion of the silt–clay banks (lower threshold of Class B3, see Table 9.3) is usually produced by flows greater than 10 m^3 s^{-1}. Out of the 11 events of this class, eight are responsible for bank slump that shows bare areas, bevelled along the contact between the crusted surface and the unweathered sediments. The volume of denudation due to these slumps is particularly large in torrential flows (approximately 55% of the total erosion of the banks). The geomorphological effectiveness of processes of this type in other ephemeral channels in the Campo de Cartagena was analysed by Conesa García (1989), giving similar results.

In the high-flow regime, and situations of prolonged bankfull, the plasticity that these sediments acquire accelerates the failure process by liquefaction causing surface mass failure, along a slightly curved plane (classes B1 and B2). Deep breaks are improbable in these banks, due to the increase in strength relative to shear stress with depth. In addition, the weakening of the bank due to erosion processes may reduce the friction angle to less than the slope. This occurs commonly when, after repeated falls of material, a small debris cone of low angle accumulates at the base of the slope. These deposits, very common in the sections studied, have a granular structure which is less cohesive than the original material.

In spite of the low sinuosity of the channel (SI = 1.17), the bank erosion rate corresponding to extreme events is considerable (−2.4 m^2 for a total of seven events of classes B1 and B2) (Table 9.4). Wide lateral bars form and divert the flow, producing local erosion of the bank. These bars present an upwards coarsening sequence, especially in the head zone, where they consist of silts and sands, sands

TABLE 9.3 Classes of morphological adjustments in the Benipila *Rambla*. Pozo de los Palos and Los Segados sections

Class B1	Overbank flow and major changes to the floodplain. Collapse of the embankments. Important reworking of sediments in transverse bars. High bedload.
Class B2	Advance of the avalanche faces of the gravel bars (riffles). Moderate bank erosion. Dissection of low terraces (1.5 m) and lateral bars. Removal of debris-flows. Median size of the coarse sediment moved: 6–10 cm. High suspended load.
Class B3	Basal erosion of the silt–clay bank and shallow planar collapses. Modification of the distances between riffles and pools. Slight removal of slag-type deposits and debris. Movement of gravels and small-sized cobbles (median size <6 cm).
Class B4	Slight movement of fine sediments (silt and sands) in the bed of the main channel. No changes are observed in the bedforms.

TABLE 9.4 Balance of erosion–sedimentation for various event classes in several channel cross-sections. Nagalte *Rambla* (1986–1991) and Benipila *Rambla* (1985–1991)

(a) Nogalte stream. Braided channel

Dates	Event occurrence			Stable bank section, 2 km upstream of Puerto Lumbreras			Eroded bank section, 1 km downstream of Puerto Lumbreras		
	N1	N2	N3	Eroded bank (m²)	Net channel change (m²)	Ratio b/ch (*)	Eroded bank (m²)	Net channel change (m²)	Ratio b/ch (*)
7 Mar. 86/7 Oct. 86	0	1	3	−0.22	+0.48	6.41	−1.04	−0.87	6.15
4 Oct. 87/10 Nov. 88	0	0	1	0.00	+0.23	6.45	−0.06	+0.12	6.03
8 Jan. 89/30 Nov. 89	1	1	2	−0.53	−0.46	5.85	−1.82	−1.45	5.60
2 Mar. 90/20 Dec. 90	0	0	1	−0.05	−0.15	5.80	−0.17	+0.09	5.95
22 Jan. 91/2 June 91	0	1	2	−0.07	+0.10	6.12	−0.66	−0.33	5.51
Total period	1	3	9	−0.97	+0.20		−3.75	−2.44	
Net bed change					+1.17			+1.31	

(b) Benipila stream. Low sinuosity channel

TABLE 9.4 (continued)

	Event occurrence			Irregular bed and unstable bank sections					
				Los Segados section			Pozo de los Palos section		
Dates	B1	B2	B3	Eroded bank (m²)	Net channel change (m²)	Ratio b/ch (*)	Eroded bank (m²)	Net channel change (m²)	Ratio b/ch (*)
20 Feb. 85/17 Nov. 86	1	3	5	−1.05	−0.78	4.48	−1.34	−0.97	2.69
28 Aug. 87/22 Nov. 88	0	0	1	−0.03	−0.17	4.40	−0.08	+0.13	2.83
15 Jan. 89/3 Dec. 89	0	1	2	−0.36	−0.24	3.89	−0.29	−0.50	3.05
11 Jan. 90/20 Dec. 90	0	0	2	−0.07	+0.10	3.95	−0.13	−0.22	3.13
22 Jan. 91/10 Dec. 91	1	1	1	−0.49	−0.31	4.11	−0.53	−0.38	3.02
Total period	2	5	11	−2.00	−1.40		−2.37	−2.07	
Net bed change					+0.60			+0.33	

(*) Relation between the bed–bars segments and the accumulated width of the interior channels

and gravels alternating with fine silt layers, and sands, gravels and cobbles. A fine sediment cover is developed on the angular coarse material of the platform to which accretion from the two floods that occurred during the period of monitoring have contributed (class B1).

The head of these bars is generally abrupt, with large-scale cross-stratification. The events of class B2 include the advance of the avalanche face and, in the case of gradual hydrograph recession, the supply of abundant fine material to the bar-tail.

At the contact between the gravel bank and the clay banks of the channel, even in slightly convex sections, the flow forms a secondary channel between the bar and the bank. Near the Caserío Pozo de los Palos, in the section that represents the less sinuous channel, the retreat of the bank has been such, that this bar loses its wedge character and the secondary channel becomes the main one.

Downstream, slag and debris deposits, which act like the hydraulic resistance of small riffles, have favoured the development of marginal bars, whose formation process varies little from the model described by Leopold (1982). The material is eroded from the sectors of flow concentration, close to one of the banks and deposited diagonally downstream in flow-divergent areas, due to the regular oscillations of the flow and the related secondary circulation. The riffles at the points of contact with the bar show significant accumulation of coarse sediments, distinguished from the avalanche face by their lack of organization and continuity.

The available data indicate a threshold for class B4 (mobilization of sands and gravels) slightly lower than that of the Nogalte stream. But in both cases, modification of bed morphology does not begin until the next threshold (point accretion and segregation of small alluvial bars, or incipient scour processes).

The two sections studied in the lower reach of the Segura River differ in size and geometry, in spite of their both having slightly sinuous planforms in wide alluvial plains. The first (Beniel–El Arenal) displays critical thresholds that are lower than those of the Algorfa section, due to its smaller bankfull capacity (95 m^3 s^{-1}). Recently, the road surface that goes along the embankment has been raised (the capacity being increased to 125 m^3 s^{-1}). In contrast, in the Algorfa section, bankfill is about 166 m^3 s^{-1}, implying a much higher threshold for class SA1 (see Table 9.5).

TABLE 9.5 Classes of morphological events in the Lower Segura channel

Beniel–El Arenal	Algorfa
Class SB1: Overflow, with important fine sedimentation in the floodplain. Collapsing of embankments and spoil banks.	*Class SA1*: Bankfull with local overflow, affecting to the floodplain.
Class SB2: Erosion of the lower part of the bank (up to a height of 2 m) and of small central and lateral bars.	*Class SA2*: Erosion of the lower part of the channel (up to a height of 2.5 m). Frequent bank collapse.
Class SB3: Modification of simple bars partially fixed by the accumulation of canes, trunks and weeds.	*Class SA3*: Slight erosion produced in the low flow levels(<1.2 m high). Slight deposition of fine material in the inner bank of the meander.

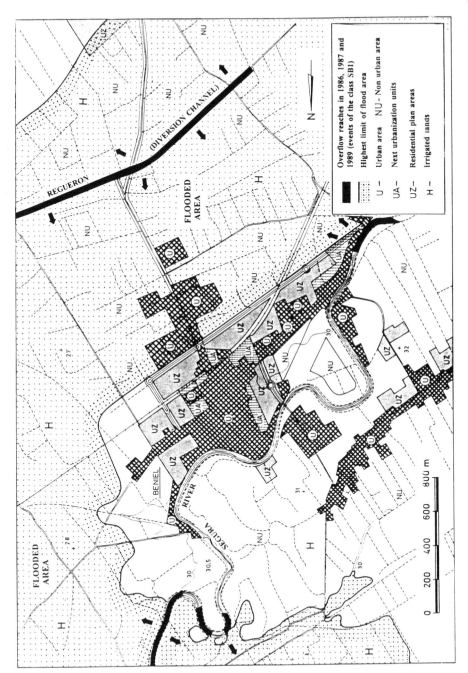

FIGURE 9.8 The Segura River in Beniel and the limits of the maximum area of flood in October 1986, November 1987 and September 1989

These are muddy meandering rivers, whose sediment load mainly consists of fine sands, silts and clays. The dominant bedforms are ripples and small long bars, attached to both banks. Basal erosion occurs in the outer bank (corresponding to 62% of events of class SB2) and lateral silt and sand deposits with epsilon cross-stratification develop at the opposing bank. The bedforms are comparable to those of facies F_1 and F_2 of Jackson (1981), although they occur before the lower end of the meander bend.

DYNAMIC-EQUILIBRIUM CONDITIONS

The response of Nogalte system to the sequence of torrential events occurring during the field period is illustrated by the morphological change observed at a site approximately 1 km downstream from Puerto Lumbreras (Figure 9.7). Detailed surveys of the channel-bed and gravel-bar morphology were carried out in March 1986 and September 1989. After the flood of March 1986, one or two main channels and several secondary channels are evident. Their orientation is controlled more by the alluvial terraces than by the perimeter of the actual *rambla*. In the centre of the channel, a transverse bar of gravel was dissected, forming an avalanche face downstream. At the lower end of the bend, close to the higher terrace, a sand-flat some 400 m long formed. In 1987 and 1988 there was a dry period during which little channel modification took place. It was not until November 1989 that any general changes in the braided system (class 1 events) were observed: the smaller channels showed more complex interweaving, the central and longitudinal bars began to overlap, the lower terrace was partially destroyed, and on the sand-flat adjoining the left bank, small debris cones were formed by lateral tributaries (gullies and *ramblizos*). An N2 event and three N3 events during the period 1990–1991 were sufficient to remove bedload, but caused a net channel change of little consequence (-0.05 m^2 and -0.24 m^2 in the analysed sections, upstream and downstream from Puerto Lumbreras respectively).

Although there was considerable lateral erosion, and sand-flats receded at several points over this six-year period (1986–1991), the channel shows a balance slightly in favour of deposition. This eroded bank-section is very close to the intersection of a broad alluvial fan, consisting of laminated sands. In general, the channel maintains a dynamic equilibrium, even though subsidence of the tectonic graben of the Lower Guadalentin, through which it flows (Conesa García *et al.*, 1991), continues. This is in contrast to the Benipila *Rambla*, where bed stability is apparent.

For the period 1985–1991, the cross-sections studied in the Benipila stream show a differential morphological response, the apparently stable bed profile contrasts with the great instability of the banks (Table 9.4). The whole fluvial system is controlled by major events of class B1 and by their relationships to the moderate events (class B2 and B3), particularly to removal of material from alluvial bars and to basal erosion of banks. Despite the removal of the gravel bars, net changes in bed topography are of small significance. Normally, when floods occur, at the base of these sections, erosion processes dominate at first, and later, as the velocity of flow declines rapidly, the deposition of suspended load follows. The result is a final balance with little change from the initial state ($+0.3$ to $+0.66$ m^2 for the same period, 1985–1991).

Bank erosion processes, however, are very important. Large amounts of sediment in suspension are carried downstream immediately after a flood, and comes mainly from the scour and slumping of loose material found along the banks. The result is a progressive widening of the channel, which takes on the shape of a trough much wider than it is deep. The frequency of the events responsible for this type of process (flooding and even moderate discharge) was greatest in the period 1985–1986. Over these years the fluvial system maintained a dynamic-equilibrium form. Debris flow removal and a higher erosion rate took place as a result of the major event occurring in February 1985. The eroding embankments remained fully active, with collapses in many places. In the more moderate events, as the bed begins to aggrade and bank widening slows, accretion on the inside of the bend plays an increasing role in channel-bank processes. From August 1987 to November 1988, no significant flooding took place. It was a period of drought, characterized by very poor vegetation cover, during which the channel morphology remained constant. During 1989 and 1990 only localized changes took place within the channel, returning it almost to its earlier form, just before the start of the monitoring period. Convex bank accretion and stabilizing woody vegetation functioned to restore natural bank stability. The trend from dynamic equilibrium towards stabilization, initiated in 1987, became especially pronounced in these years. In 1991 there was overbank flow, causing scour rather than deposition on the floodplain.

Finally, downstream on Segura River, at monitored sites (Beniel and Algorfa – "Rojales"), the changes in the fluvial system suggest that class SB1 and SA1 events (Figure 9.8) tend to be associated with vertical aggradation processes on the floodplain, and class SB2 and SA2 events with basal erosion of the channel. In minor

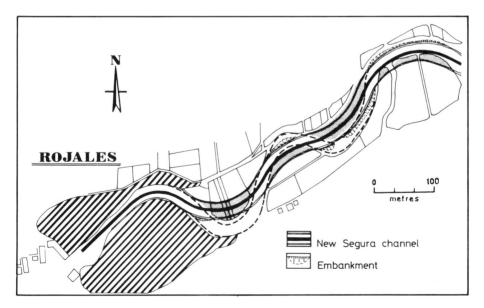

FIGURE 9.9 Project for the canalization of the Segura River. Rojales–Guardamar sector (Alicante)

events, and during declining discharge after the flood peak in major events, a sediment sequence develops in which the size of individual particles increases within each layer. In all classes, sedimentation processes are predominant. Currently, both sites are included in an ambitious programme of canalization (Figure 9.9) that will mitigate considerably the risk of floods in the whole Lower Segura Valley.

Prior to the period of study, the major event threshold defined in the present work was exceeded (e.g. 2000 m^3 s^{-1} in October 1973, in Nogalte; >500 m^3 s^{-1} in September 1919 and October 1948 in Benipila). These catastrophic events are rare, but constitute high-magnitude floods which affect the overall morphology of the fluvial system within the long-term dynamic equilibrium.

CONCLUSIONS

The definition of thresholds for the different classes of morphological adjustments in ephemeral channels of south-eastern Spain requires a knowledge of the environmental conditions of this semi-arid region and, in particular, the response of the fluvial systems to the irregular and torrential regime of rains. The higher thresholds correspond to extreme events (85–166 m^3 s^{-1}), causing an overall modification of the system. Channel overflow (class 1) events occur with a recurrence interval of 1.2 to 6 years. In the Lower Segura River, the supply of silts to the floodplain provide fertility, but the immediate effects are harmful to the rural population. Collapse of the embankments does not cause important modifications of the depositional units. In the Nogalte *Rambla*, classes N1 and N2 mean considerable changes in sand-flat and mud-sand accumulation in the low and medium alluvial terraces. On the other hand, the discharges that exceed the threshold of class SB1 normally form crevasse-splays of sands and gravels on the lower terraces.

The events of class 2, related to floodpeak discharges of 27 to 43 m^3 s^{-1}, cause significant modifications within the channel depending on the geometry and lithology of the cross-sections. These changes affect the position, size and texture of the bedforms and, in some sections, the perimeter of the main channel. The major morphological adjustments within this group occur in the base of the outer banks. Frequently, the low coherence of the bank materials accentuates the degree of instability, setting the minor threshold of the class around 1 m flow depth and velocities greater than 0.8 m s^{-1}. This type of event has a recurrence interval of 1.4 years for sections of *rambla* and 3.5 to 4 months for the lower reaches of the Segura.

The periods of analysis, 1987–1988 and 1990 corresponded with drought and stability. With the major events in 1989 (Nogalte) and 1991 (Benipila) this stability was disturbed and the system returned to a pattern of episodic adjustment.

To a large extent, the morphological features of ephemeral channels and the degree of geomorphic response to torrential discharges depend on the relative importance of scour and deposition processes during the major events. These, in turn, depend on the environmental characteristics of the basins, the subsequent discharge regimes of both water and solids, channel geometry and, in some cases, the effect of external factors such as base-level changes due to subsidence. An example of this external effect can be observed in the Nogalte stream, whose braided channel maintains a dynamic equilibrium, in spite of the presence of an active tectonic border to the north of the Alto Guadalentin graben. The Benipila stream and the Lower Segura, despite the apparent stability of their beds, also show a tendency to dynamic equilibrium, this time related to the base level of the Mediterranean Sea.

Of the three areas analysed, the two that correspond to torrential streams (Nogalte and Benipila *ramblas*) seem to operate in the long run in dynamic equilibrium, the historic floods being the driving force fundamental to the understanding of the magnitude of these adjustments. On the other hand, the Lower Segura, whose regime has been deeply altered by multiple abstractions for irrigation (especially, that of El Reguerón) and by the construction of numerous upstream dams, has suffered a very substantial change in transport conditions, that make it potentially less active. Its recent canalization as far as its mouth in Guardamar, with many meander cutoffs and the subsequent increase in the flow velocity, will modify the present equilibrium once more and will vary the thresholds established in this study.

ACKNOWLEDGEMENTS

The author wishes thank Mr M. Menéndez Campo, Chief of the Hydrometry Division, and Mr L. Quintas, Chief of the Systems Engineering Division, of CEDEX Madrid, for permission to consult their gauging records. I also acknowledge the cooperation and help provided by the Meteorological Office, Guadalupe – Murcia, and by the Cartographic Service of Murcia Region.

REFERENCES

Allen, J. R. L. 1982. Sedimentary structures. Their character and physical basis. *Developments in Sedimentology*, 30B, Elsevier, Vol. 2, Ch. 2, pp. 53–100.

Bluck, B. J. 1982. Texture of gravel bars in braided streams. In: Hey, R. D., Bathurst, J. C. and Thorne, C. R. (eds), *Gravel-bed Rivers*, Wiley, Chichester, pp. 339–353.

Castelltort, F. X. and Marzo, M. 1986. Un modelo deposicional de abanicos aluviales arenosos originados por corrientes efímeras: el Muschelkalk medio de los Catalánides. *XI Congreso Español de Sedimentología*, 47.

Clark, P. B. and Davies, S. M. A. 1988. The application of regime theory to wadi channels in desert conditions. In: White, W. R. (ed.), *International Conference on River Regime*, Wiley, Chichester, pp. 67–82.

Conesa García, C. 1986. Movilidad de las barras en lechos de rambla del Sureste Peninsular (España). In: *Estudios sobre geomorfología del Sur de España*, Univ. of Murcia, Univ. of Bristol, COMTAG, International Geographical Union, Murcia.

Conesa García, C. 1989. La acción erosiva de las aguas superficiales en el Campo de Cartagena, Univ. of Murcia, CajaMurcia, 137 pp.

Conesa García, C., Sánchez Medrano, R., Solis García-Barbón, L. and Cabezas-Rubio, F. 1991. Aplicación de métodos de prospección geofísica a la evolución de formas de drenaje y facies sedimentarias del Cuaternario en el Valle Alto del Guadalentín. *VIII Reunión Nacional sobre Cuaternario*, AEQUA, Valencia, 24.

Dury, G. H. 1973. Magnitude frequency analysis and channel morphology. In Morisawa, M. (ed.), *Fluvial Geomorphology*, Publications in Geomorphology, Binghamton, pp. 91–122.

Einstein, H. A. and Shen, H. W. 1964. A study of meandering in straight alluvial channels. *Journal of Geophysical Research*, **69**, 5239–5247.

Glennie, K. W. 1970. Desert sedimentary environments. *Developments in Sedimentology*, **14**, 222 pp.

Harvey, A. M., Hitchcok, D. H. and Hugues, D. J. 1982. Event frequency and morphological adjustment of fluvial systems in Upland Britain. In: Rhodes, D. D. and Williams, G. P. (eds), *Adjustments of the Fluvial Systems*, George Allen & Unwin, London, pp. 139–168.

Jackson, R. G. 1981. Sedimentology of muddy fine-grained channel deposits in meandering stream of the American middle west. *Journal of Sedimentary Petrology*, 51, 1169–1192.

Lekach, J. and Schick, A. P. 1982. Suspended sediment in desert floods in small catchments. *Israel Journal of Earth Sciences*, 31, 141–156.

Leopold, L. B. 1982. Water surface topography in river channels and implications for meander development. In: Hey, R. D., Bathurst, J. C. and Thorne, C. R. (eds), *Gravel-Bed Rivers*, Wiley, Chichester, pp. 359–388.

Leopold, L. B., Emmett, W. M. and Myrick, 1966. Channel and hillslope processes in a semi-arid area, New Mexico. *US Geological Survey Professional Paper*, **352-G**.

McKee, E. D., Crosby, E. J. and Berryhill, H. L. J. 1967. Flood deposits, Bijou Creek, Colorado, June 1965. *Journal of Sedimentary Petrology*, 37, 829–851.

Nouh, M. 1988. Regime channels of an extremely arid zone. In: White, W. R. (ed.), *International Conference on River Regime*, Wiley, Chichester, pp. 55–66.

Pickup, G. and Rieger, W. A. 1979. A conceptual model of the relationship between channel characteristics and discharge. *Earth Surface Processes*, 4, 37–42.

Rhodes, D. D. and Williams, G. P. (eds). 1982. *Adjustments of the Fluvial System*. George Allen & Unwin, London.

Richards, K. 1982. *Rivers. Form and Process in Alluvial Channels*. Methuen, London, pp. 138–145.

Schick, A. P. 1967. Suspended sampler. *Revue de Géomorphologie Dynamique*, 4, 181–182.

Tunbridge, I. P. 1984. Facies model for sandy ephemeral stream and clay complex; the Middle Devonian Irentishoe Formation of North Devon, U.K. *Sedimentology*, 31(5), 697–716.

Wolman, M. G. and Gerson, R. 1978. Relative scales of time and effectiveness of climate in watershed geomorphology. *Earth Surface Processes*, 3, 189–208.

Yang, C. T. 1983. Minimum rate of energy dissipation and river morphology. *Symposium on Erosion and Sedimentation*, Colorado State University, Fort Collins, Colorado, 3.2–3.19.

Yang, C. T. and Molinas, A. 1988. Dynamic adjustments of channel width and slope. In: White, W. R. (ed.), *International Conference on River Regime*, Wiley, Chichester, pp. 17–28.

Yang, C. T. and Song, C. S. 1982. Dynamic adjustments of alluvial channels, In: Rhodes, D. D. and Williams, G. P. (eds), *Adjustments of The Fluvial System*, George Allen and Unwin, London, pp. 55–67.

Yang, C. T. and Song, C. S. 1986. Theory of minimum energy and energy dissipation rate. *Encyclopedia of Fluid Mechanics*, vol. 1, ch. 11, Gulf Publishing Company, pp. 353–399.

10

Channel Changes on the Po River, Mantova Province, Northern Italy

D. CASTALDINI

Dipartimento di Scienze della Terra, Universitá degli Studi di Pisa, Italy

AND

S. PIACENTE

Dipartimento di Scienze della Terra, Universitá degli Studi di Modena, Italy

ABSTRACT

This research examines channel changes on the Po River in northern Italy. The study is based on detailed bibliographic research, the examination of historical documents and maps, the interpretation of aerial photographs taken in different years, and a morphological survey of the present river channel.

From the Bronze Age to the Middle Ages, the Po River flowed within a meander belt about 20 km in breadth which migrated northward, partly influenced by tectonic movements of buried geological structures.

Between the 16th and 18th centuries, the course of the Po River was quite unlike its present course.

From the 19th century to the 20th century, there was progressive narrowing of the low-water channel, and a reduction of the sinuosity and length of the river.

After the 1970s, the construction of bank protection structures markedly increased, obstructing the river's natural evolution and causing further narrowing of the low-water river channel.

It is concluded that the changes in the Po River channel observed during the 20th century are at least partially a consequence of river engineering.

INTRODUCTION

The fluvial system can be treated as either a physical system or an historical system (Schumm, 1977). This paper focuses on the evolution and changes of a section of the Po river system over historical times.

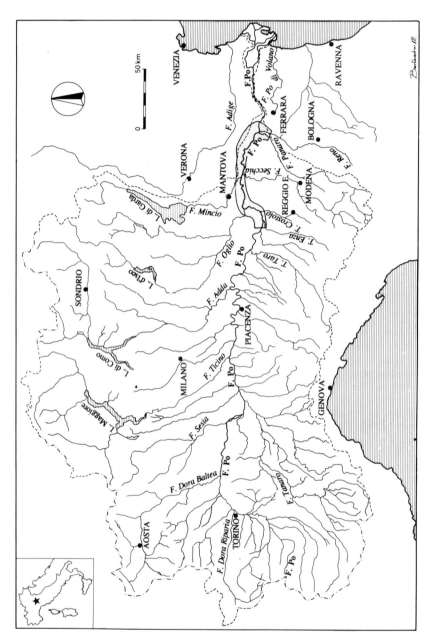

FIGURE 10.1 The Po River basin and the study area

In particular the aim of this study is twofold: (a) to document the history of channel changes on the Po River from the Bronze Age to the present; (b) to examine the contemporary morphodynamics of the Po River.

GEOGRAPHICAL AND GEOLOGICAL OUTLINE

The Po River is the longest river in Italy (673 km) and it has always been a significant geographical and socio-economic element in the region.

The tract considered in this study is about 100 km long and is situated mainly in the Province of Mantova, in the central-eastern sector of the Po Plain (northern Italy) (Figure 10.1). The Po Plain represents one of the principal morphological units of the Italian peninsula and it is the most extensive plain in Italy (approximately 46 000 km^2, representing 71% of all the plain areas and 15% of the nation's territory).

The Po basin is situated in a temperate climatic zone (Type C, Köppen's classification), or more specifically, in a subcontinental temperate climatic zone. In particular from the pluviometric viewpoint the study area corresponds to one of the least rainy sectors within the Po basin with average annual rainfall from about 600 to 700 mm, with seasonal peaks concentrated in the autumn and spring, and minimum amounts in the summer (Cati, 1981).

Mean values for the flow rate have been calculated for the study tract, near Ferrara, at 1470 m^3 s^{-1} in the period 1918–1970 (Cati, 1981).

The surficial alluvial deposits (grain size ranging from sand to clay) in the study area are of Holocene age south of the Po and of Holocene–Upper Pleistocene age to the north (Cremaschi, 1987; Panizza *et al.*, 1987).

The study area lies at the junction between the southern border of the "Pedalpine Homocline" and the northern margin of the "Ferrara Folds" (see Pieri and Groppi, 1981). The latter are a continuation of the Apennine chain in the Po Plain foredeep and the former represents the continuation, within the subsurface of the plain, of the homocline structure that characterizes the Alpine margin in the Verona area.

METHODS OF STUDY

Several sources of data were examined:

(i) bibliographic research;
(ii) research on historical documents and maps;
(iii) morphological analysis through the interpretation of aerial photographs taken in various periods;
(iv) field survey of the morphology of the present-day high-water river channel.

The analysis of historical maps in particular led to several problems regarding the compilation of the synthetic maps. Such problems were due, not only to the validity of the documents obtained, but also to difficulties involved in comparing maps with different cartographic scales.

The maps examined consist of the following:

(i) Historical maps dating from the 16th, 17th and 18th centuries (at various scales), in the collections of the Accademia Nazionale Virgiliana, the Biblioteca Comunale and mainly, the Archivio di Stato in the city of Mantova.
(ii) 19th century maps (scale: 1:10 000 approximately) from the Catasto del Regno Lombardo-Veneto (Land Register), conserved in the Archivio di Stato of Mantova.
(iii) Map compiled in 1876 (scale: 1:75 000), conserved in the Archivio di Stato of Mantova.
(iv) Istituto Geografico Militare (IGM) maps (scale: 1:25 000), 1885–1888, 1912, 1933–1935 and 1970–1973 editions of the F. 62 Mantova, F. 63 Legnago, F. 74 Reggio Emilia and F. 75 Mirandola maps.
(v) Istituto Geografico Militare (IGM) maps (scale: 1:100 000): F. 62 Mantova (updated to 1954), F. 63 Legnago (updated in 1953), F. 74 Reggio Emilia (updated in 1950), F. 75 Mirandola (updated in 1950).
(vi) Regional Technical Maps (CTR) of the Lombardy Region (scale: 1:25 000 and 1:50 000; updated in 1983), of the Veneto Region (scale: 1:10 000; updated in 1990), and of the Emilia-Romagna Region (scale: 1:25 000; updated in 1984).

The aerial photographs examined include black and white aerial photographs at the scale of 1:33 000 (approximately) taken in 1955 and color aerial photographs taken in 1981 at the scale of: 1:20 000.

EVOLUTION OF THE PO RIVER

Evolution from the Bronze Age (before the 8th century BC to the Middle Ages (15th century)

On the basis of the bibliographic research conducted, the evolution of the Po River (shown in Figure 10.2) can be summarized as follows.

After the emergence of the Po Plain, in the Upper Pleistocene, the Po River flowed much more to the south of the present course and into the Adriatic Sea southeast of Ravenna. This course is documented by Po River deposits at depths of over 100 m with respect to the present surface (Gasperi and Pellegrini, 1984).

Around the year 1000 BC, the Po River forked between Brescello and Guastalla or east of Casalmaggiore, creating a main branch ("Po di Adria") and one ("Po di Spina") or more minor branches parallel to it, along a strip bound on the south by the cities and towns of Novi, Concordia, Bondeno and Ferrara.

Near the beginning of the 8th century BC, the Po breached in the area between Brescello and Guastalla, heading north and joining the present-day "Po Vecchio" canal (which means the "Old Po") as far as Sustinente. From Sustinente as far as Bergantino, the course of the river was not too far from that at present.

In about the same period (9th to 8th century BC), the Po River breached in the vicinity of Sermide due to the overflowing of the "Po di Adria" and from Sermide, it flowed towards Bondeno and then to Ferrara ("Po di Ferrara").

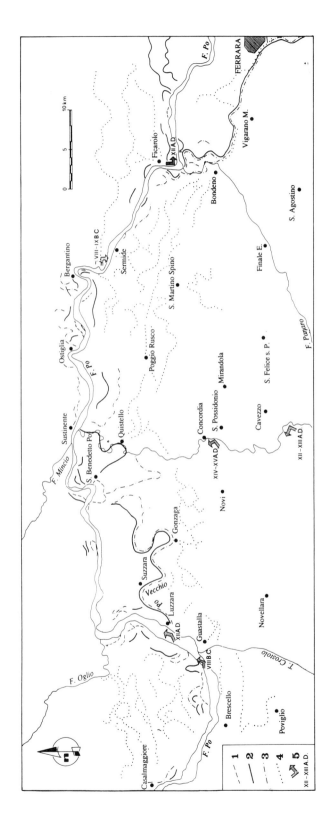

FIGURE 10.2 Evolution of the Po River from the Bronze age to the Middle Ages (from Castaldini, 1989; simplified). 1 – Paleo-river of the modern age. 2 – Paleo-river of the Middle Ages. 3 – Paleo-river of the Roman period. 4 – Paleo-river of the Bronze Age. 5 – Main river diversion with indication as to age (AD = Anno Domini, BC = Before Christ)

In the Roman period and the early Middle Ages the Po still followed the course of the "Po Vecchio". In the early Middle Ages, however, the river in the western side tended to shift northward, creating a branch further north, first called "Padus Lirone" and then simply "Padus". East of S. Benedetto Po, the Po flowed in a single channel that was not very different from the present-day river channel as far as Ficarolo and, from there, it continued towards Bondeno and Ferrara.

Around 1150–1200 two very important diversions occurred near Luzzara and near Ficarolo. More specifically, because of the "Luzzara breach", the river abandoned the "Po Vecchio", and because of the "Ficarolo breach", the Po continued eastward, abandoning the course towards Ferrara.

At the close of the 15th century, the course of the Po River was similar to its present-day course.

The migration of the Po northward was conditioned by the tectonic activity of the buried geological structures (Panizza et al., 1987), subsidence, variations in sea level and by the hydrologic regime of the river.

In the course of the last five centuries, the Po has been forced to flow along a course controlled by engineers, that is, within a strip of land bounded by artificial levees. Of the various controls on fluvial behaviour, the one that has most influenced the evolution of the channel has been the hydrologic regime fluctuations attributable mainly to climatic factors.

Evolution from the 16th to the 18th century

Documentation of the course of the Po in this period consists of numerous historical maps representing the Po Plain sector considered. It should be kept in mind, however that the documents available for the 16th century are not very reliable (Ferrari, 1985). The most significant maps used for the reconstruction of the Po evolution in the period from the 17th to the 18th century are briefly described below.

The first document utilized is the Magini (1620) approximately 1:300 000 map (Figure 10.3) because it is based on reliable data and was compiled using scientific criteria (Ferrari, 1985). In fact the map is orientated, showing latitude and longitude degrees as well as a detailed drainage network.

Most of the later maps were directly derived from the Magini (1620) map or are slavish reproductions of it. Thus, there are no historical maps of significant value until the approximately 1:200 000 map by Coronelli (1690).

The maps dating from the first half of the 1800s are still mainly derived from the map by Magini (1620). Therefore, the reconstruction of the course of the Po River for the above-mentioned period was compiled on the basis of the approximately 1:200 000 map printed by the Officine Homann (1735).

Numerous maps dating from the close of the 18th century were directly derived from maps compiled in the first half of the century. Therefore, the reconstruction of the drainage network in the second half of the 18th century is based on the approximately 1:250 000 map by Boselli (1801) which reflects the geographic situation in 1796.

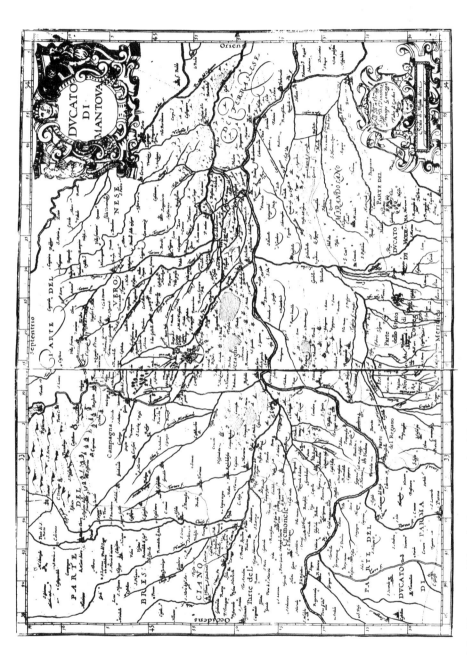

FIGURE 10.3 Example of historical map examined for the reconstruction of the Po evolution from the 17th to the 18th century (Magini, 1620 map. Original scale 1:300 000 approx. Collection of the Archivio di Stato of Mantova)

The evolution of the Po River in the 17th and 18th centuries (Figure 10.4) was reconstructed from the analysis of the maps listed above. The analysis permitted the identification of the formation of a large meander west of S Benedetto Po at the beginning of the 18th century (in fact, the 17th-century maps show a rectilinear course of the Po River in the same tract). It was noted that between Luzzara and the confluence with the Oglio River, the Po River was characterized by a progressive east–west shifting of the channel whereas the section downstream from the confluence with the Secchia River had a configuration similar to that at the present,

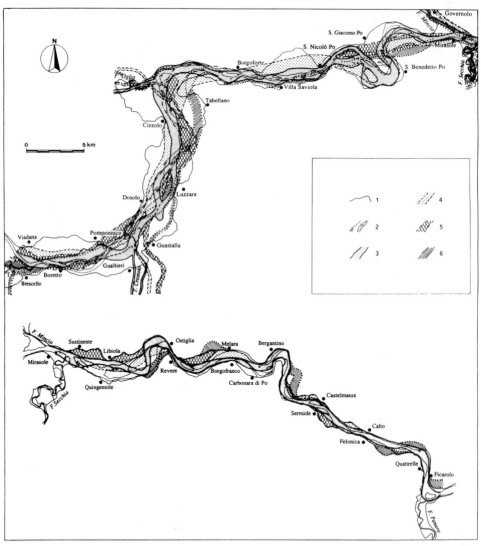

FIGURE 10.4 Evolution of the Po River from 17th to the 18th century. 1 – Main leeves (1983). 2 – River channel in 1983. 3 – River channel at the end of the 18th century. 4 – River channel in the first half of the 18th century. 5 – River channel at the end of the 17th century. 6 – River channel at the beginning of the 17th century

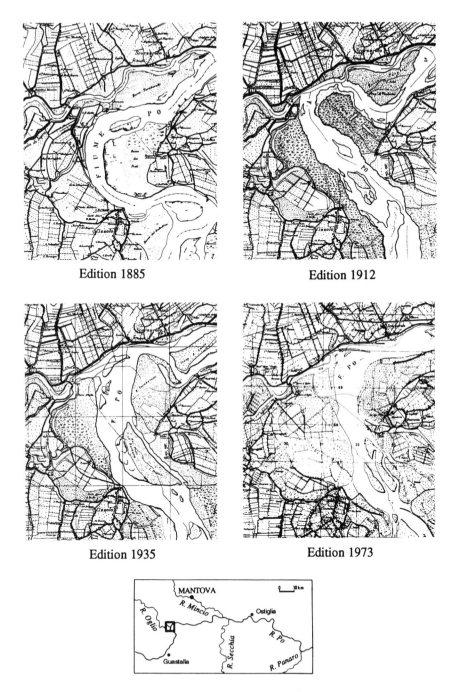

FIGURE 10.5 Example of Istituto Geografico Militare (IGM) maps examined for reconstruction of the Po evolution from the 19th to the 20th century. Gazzuolo map, original scale 1:25 000. Reproduced by permission of Archivio di Stato di Mantova

as early as the close of the 18th century. Moreover, it was observed that the islands, which are virtually absent on the 17th-century maps, are numerous on the 18th-century maps (it is not clear, however, whether this situation is due to fluvial morphodynamics or to a change in mapping criteria).

Evolution of the Po River from the 19th to the 20th century

The evolution of the Po in this period was reconstructed on the basis of the 19th to 20th-century maps (Figure 10.5) listed above in the "Methods of study" section.

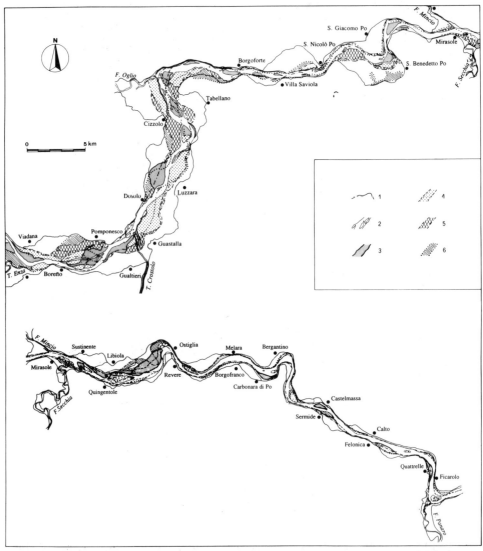

FIGURE 10.6 Evolution of the Po River from the 19th to the 20th century. 1 – Main levees (1983). 2 – River channel in 1983. 3 – River channel in the second half of the 20th century. 4 – River channel in the first half of the 20th century. 5 – River channel in the second half of the 19th century. 6 – River channel in the first half of the 19th century

The analysis and comparison of the individual maps representing the various periods led to the following observations (Figure 10.6). The wide meander west of S. Benedetto Po was abandoned due to an artificial incision, made between the end of the 18th century and the beginning of the 19th century. In the 20th century there was a progressive narrowing of the low-water river channel that is more marked in the western section; a gradual reduction of the length of the meandering reach of the river; and a reduction of the length of the course of the river by a few kilometres (3 km approximately). Human modification of the river is evident in many locations. The first groynes appear on the maps dating from 1933–1935, but only in the section upstream from the Mincio River. Their main function was to render the river navigable during the periods of low water, even though they also obviously functioned as bank protection. From the 1970s, the number of groynes increased greatly and today the river channel is highly influenced by human activities. In addition to groynes, rock-fill protection systems for the banks have been added.

PRESENT-DAY MORPHODYNAMICS

A detailed river morphology map (work scale 1:25 000) has been compiled for the high-water channel between the main levees of the Po River on the basis of aerial photograph interpretation and field surveys. The river morphology map (Figure 10.7)

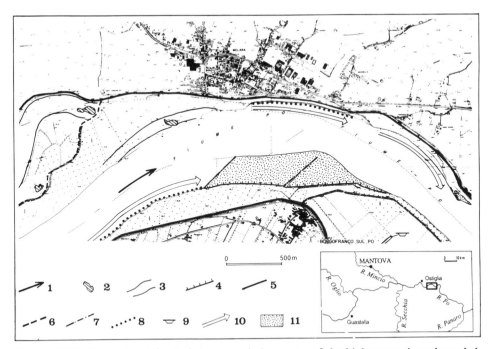

FIGURE 10.7 Example of detailed river morphology map of the high-water river channel. 1 – River flow. 2 – Small body of water. 3 – Paleo-river channel at the high-water channel level. 4 – Scarp. 5 – Main levee. 6 – Secondary levee. 7 – Groyne. 8 – Rock-fill protection. 9 – Abandoned quarry (dry floor). 10 – Bank tract undergoing erosion. 11 – Present-day or recent sedimentation area

includes: data on hydrography, fluvial landforms (paleo-river channels and scarps), forms and structures of human origin (levees, bank protection, quarries and sifting plants), and the areas undergoing erosion and sedimentation at present.

Among the elements pertaining to the hydrography, there are numerous small bodies of water indicated on the river morphology map of the study tract (Figure 10.8). They are of limited dimensions and are often characteristic of paleo-river channels.

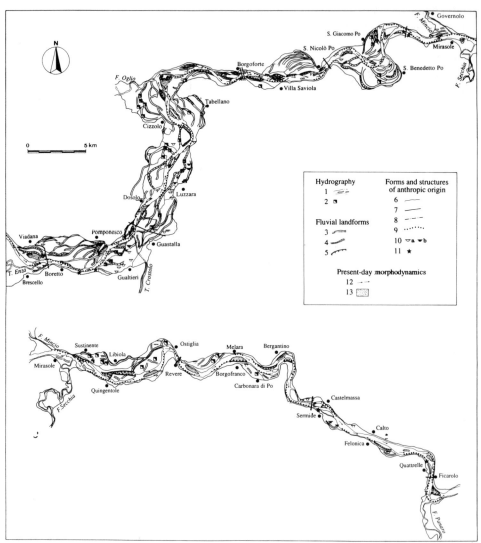

FIGURE 10.8 River morphology map of the study tract. 1 – River channel in 1983. 2 – Small body of water. 3 – Paleo-river channel at the high-water channel level. 4 – Paleo-river channel confined within the high water channel level. 5 – Scarp. 6 – Levee. 7 – Bridge. 8 – Groyne. 9 – Rock-fill protection. 10 – Abandoned quarry: (a) dry floor; (b) flooded floor. 11 – Sifting plant. 12 – Bank tract undergoing erosion. 13 – Present-day or recent sedimentation area

Paleo-river channels are the most typical of the fluvial landforms. They are either at the high-water channel level or marked by scarps of about one metre in height. Numerous paleo-river channels have been identified in the Viadana–Ostiglia tract where the high-water channel area widens (2 to 3 km). Traces of the meander in the vicinity of S. Benedetto Po, which was active in the 18th century, are particularly evident.

Other characteristic fluvial landforms are scarps and they are distributed almost everywhere. For the most part, as mentioned above, they are characteristic of the paleo-river channels confined within the high-water plain area. In other cases, when they are closer to the low-water channel, they separate the high-water channel level from the areas of present-day or recent sedimentation.

Forms and structures of human origin are widely distributed; the river morphology map shows the system of main levees and secondary levees (Figure 10.7). The former reach heights of over 5–6 m above the surrounding plain level, whereas the latter, which are located within the high-water channel area, are relatively low, generally 3–4 m in height, but in many cases, only 1–2 m. Other man-made features that condition the present-day evolution of the Po River channel are represented by the bank protection structures: rock-fill protections and groynes. These structures are situated along the concave banks of the river and the tracts of protected river banks vary in length from several hundred metres to several kilometres. In general, the tracts of banks protected by rock-fill, usually only a few metres high, are followed by groynes.

Quarries, all of them abandoned, are numerous; they have been differentiated into two groups: those that have flooded floors and those that have dry floors. Moreover, areas with sifting plants have been indicated because they represent areas modified by human activity and because they generally indicate present-day river channel excavations.

Bank tracts undergoing erosion and present-day or recent sedimentation areas have been indicated. It should be noted that the latter term is defined as those mainly sandy areas free of vegetation observed in the aerial photographs and/or at the time of the field survey.

The erosional capacity of the Po River obviously affects the concave banks. Erosion along the banks protected by rock-fill has led to the removal of the rock-fill but without creating noticeable bank recession. Along the bank tracts unprotected by rock-fill, erosion undercuts the banks causing bank recession. For some tracts, bank recession of about 10 m during 10 years has occurred.

The recent and present-day sedimentation areas have developed mainly along the convex banks and downstream from the groynes (Figure 10.7).

The field survey revealed a generalized and very recent increase in the deposition areas. In several cases, deposition downstream from the groynes has led to the joining of islands to the bank during the low-water periods.

CONCLUSIONS

Bibliographic research revealed that in the earliest period considered, from the Bronze Age to the Middle Ages, the Po River flowed in a belt about 20 km wide, between the Oglio outlet–Bergantino and Poviglio–Bondeno alignments. A position of the river that differs little from its present position was established after significant river diversions in the 8th century BC in the vicinity of Guastalla and Sermide, and in the 12th century AD, near Luzzara and Ficarolo. This shift northward was also conditioned by tectonic movements of buried geological structures ("Ferrara Folds").

In the period from the 17th to the 18th century AD, the Po already followed a course similar to that at present with a meander belt that widened upstream from the Oglio outlet. The formation of a large meander west of S. Benedetto Po in the 18th century is worthy of note. It was to remain active for almost an entire century.

The evolution of the river in more recent periods (19th–20th century) is characterized by progressive narrowing of the low-water river channel that is more accentuated in the western section and a reduction in the winding course and overall length of the river.

Bank protection works, starting from the 1930s, markedly increased from the 1970s and have now given the river channel a distinctly engineered appearance.

The analysis of present-day morphology revealed that erosion along the concave banks has not caused a consistent recession of the banks, except in some places, because extensive tracts of the banks are protected by defence structures. Along the convex banks and downstream from the numerous groynes, deposition of sediments has led to the advance of the banks themselves and, in some cases, to the joining of the islands to the bank. These factors hinder the natural evolution of the river and lead to a progressive narrowing of the low-water channel. Therefore, it may be concluded that the progressive narrowing of the low-water channel, the reduction of its winding course and overall length, observed during the 20th century, can be correlated to some extent with the increase in human intervention in the river.

ACKNOWLEDGEMENTS

This study was conducted in the frame of the Italian National Project Catene montuose e pianure: geomorfologia strutturale ed evoluzione in Italia e in altre aree mediterranee". Financial support was provided by the Ministero dell'Università e Ricerca Scientifica e Tecnologica (MURST) grants 40% (Coordinator of the Modena University, prof. Mario Panizza).

REFERENCES

Boselli, G. 1801. Carta topografica delle città e del territorio di Mantova coi paesi confinanti nello stato suo politico sul principio dell'anno 1796. Historical map, approx. scale 1:250 000, Archivio di Stato, Mantova.

Castaldini, D. 1989. Evoluzione della rete idrografica centro padana in epoca protostorica e storica. In: *Atti Conv. "Insediamenti e viabilità nell'alto ferrarese dall'età romana al Medioevo"*, Cento, pp. 115–134.

Castaldini, D. and Piacente, S. 1991. Evoluzione storica e morfodinamica attuale del fiume Po nel tratto tra Viadana e S. Benedetto Po (Provincia di Mantova). *Rivista Geografica Italiana*, **98**, 345–379.

Cati, L. 1981. *Idrografia e Idrologia del Po*. Ufficio Idrografico Po, Pubblicazione, n. 18, Ministero Lavori Pubblici, Servizio Idrografico, Roma, 310 pp.

Coronelli, V. 1690. Il Ducato di Mantova nella Lombardia. Historical map, approx. scale 1:200 000, Accademia Nazionale Virgiliana, Mantova.

Cremaschi, M. 1987. *Paleosols and Vetusols in the Central Po Plain (Northern Italy): a Study in Quaternary Geology and Soil Development*, Unicopli, Milano, 306 pp.

Ferrari, D. (ed.) 1985. *Mantova nelle stampe*. Grafo, Brescia, 210 pp.

Gasperi, G. and Pellegrini, M. 1984. Strutture geologiche ed idrografia della bassa Pianura Modenese. In: *Atti Conv. "Mirandola e le terre del basso corso del Secchia"*, Deputazione Storia Patria Antiche Province Modenesi, 76, Modena, pp. 3–20.

Istituto Geografico Militare (IGM) 1885–1973. Carta Topografica d'Italia alla scala 1:25 000 e 1:100 000 dei F. 62 Mantova, F. 63 Legnago, F. 74 Reggio Emilia, F. 75 Mirandola, IGM, Firenze.

Magini, G. A. 1620. Ducato di Mantova. Historical map, approximate scale 1:300 000, Biblioteca Comunale, Mantova.

Officine Homanng 1735. Ducatus Mantuani, ceu sedis belli 1733–35 recentissima delineatio, Historical map, approximate scale 1:200 000, Archivio di Stato, Mantova.

Panizza, M., Castaldini, D., Cremaschi, M., Gasperi, G. and Pellegrini, M. 1987. Ricostruzione paleogeografica e geodinamica tardo-pleistocenica ed olocenica dell'area centropadana tra Verona e Modena. In: Ente Nazionale Energia Elettrica (ENEL), *Contributi di preminente interesse scientifi o agli studi di localizzazione di impianti nucleari in Piemonte e Lombardia*, vol. 2, Roma, 410 pp.

Pieri, M. and Groppi, G. 1981. Subsurface geological structure of the Po Plain, Italy, Consiglio Nationale delle Ricerche, pubbl. 414 Progetto Finalizzato Geodinamica, Roma, 13 pp.

Regione Emilia-Romagna 1984. Carta Tecnica Regionale a scala 1:25 000, Regional Technical Map (CTR), Bologna.

Regione Lombardia 1983. Carta Tecnica Regionale a scala 1:25 000 e 1:50 000, Regional Technical Map (CTR), Milano.

Regione del Veneto 1990. Carta Tecnica Regionale a scala 1:10 000, Regional Technical Map (CTR), Venezia.

Schumm, S. A. 1977. *The Fluvial System*. Wiley, Interscience, New York.

11

The Impact of River Regulation, 1850–1990, on the Channel and Floodplain of the Upper Vistula River, Southern Poland

A. LAJCZAK

Research Centre for Nature Protection, Polish Academy of Sciences, Cracow, Poland

INTRODUCTION

In many places during the past hundred or even thousand years, river channels and valley-floor morphology have been affected by human activity (Hickin, 1983; Gregory, 1987). Changes in river regimes and rates of sediment transport have followed the deforestation and intensification of land use in river catchments; more recent causes have been mining, the initiation of urbanization, and the extraction of channel materials for industry (Hammer, 1972; Graf, 1975; Park, 1977; Castiglioni and Pellegrini, 1981; Gregory and Madew, 1982; Starkel, 1983; Gregory and Brookes, 1983; Klimek, 1987a, 1987b: Szumanski and Starkel, 1990). The impact of human activity upon discharge, including increased peak flows, has caused a variety of channel changes (Wolman, 1967; Park, 1977; Brookes, 1990). These human-induced changes also have caused rapid increases in sediment transport and sedimentation rates (Strahler, 1956; Wolman, 1967; Trimble, 1970, 1976; Sundborg, 1983, 1986; Gregory, 1987; Klimek, 1987a, 1988; Knox, 1987; Starkel, 1987; Dai, 1988; Knighton, 1989, 1991), and have been superimposed on other river adjustments, such as those induced by changes in climate (Schumm, 1968). Opposing geomorphic effects are caused by dams, which regulate the river flow and trap as much as 100% of incoming material (Gong, 1987; Lajczak, 1994a, 1994b).

One of the reasons for the acceleration of the changes in channel and floodplain morphology is channelization. These changes are most pronounced immediately after river confinement (Lajczak, 1994c, 1994d). Direct and indirect effects of river channel regulation which have occurred since the beginning of the 19th century include significant changes in channel pattern and geometry resulting from shortening, narrowing and deepening of channels (Brookes, 1990; Wyzga, 1993).

River Geomorphology. Edited by Edward J. Hickin
© 1995 John Wiley & Sons Ltd

210 River Geomorphology

These significant changes in river hydraulics have caused rapid river channel incision and rapid increases of sediment transport rates and have led to the modification of floodplain geomorphology (Lajczak, 1994c, 1994d).

This paper aims to provide a quantitative analysis of the causes and effects of

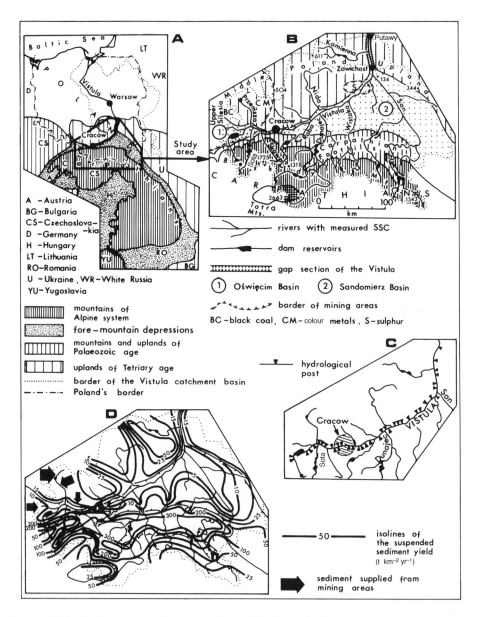

FIGURE 11.1 Location map. A – the piedmont Vistula on the background of Poland and the Carpathians; B – the main morphological units in the river catchment (SSC: suspended-sediment concentration; colour metals: Ag, Pb, Zn); C – gauging stations on the piedmont Vistula used in analysis; D – the rates of the suspended sediment yield

changes in the channel geometry and hydraulics of the Upper Vistula River and of the channel and floodplain remodelling caused by the river channelization introduced after the 1840s. These changes are examined against the background of changes in fluvial processes on the Vistula apparent since the start of human impact on the river. Figure 11.1 shows the location of the river reach studied.

STUDY AREA

The Vistula River, the longest in Poland, has a length of 1047 km, a drainage area of 194 424 km^2, and a mean discharge at its mouth of 1250 m^3 s^{-1}. The upper reach of the river, with a length of about 400 km and a drainage area of over 50 000 km^2, is located within the fore-mountain lowland between the Western Carpathians and the Middle Poland Upland (Figure 11.1B). This river reach could be termed "the *piedmont* Vistula" although this term is not quite equivalent to the term "*fore-mountain* Vistula".

The portion of the Vistula analysed in this paper drains the heavily denuded Carpathians, the mining areas and the loess regions. It is a river which has undergone dramatic and very rapid changes in channel and floodplain morphology following the initiation of channelization.

Precipitation in the area ranges from 500–600 mm in the lowland fore-mountain area to 2000 mm in the high mountains. The area receives 50–70% of its annual precipitation in the form of rain, especially during the period May–August. The hydrologic regime of this part of the Vistula is governed by a combination of highly variable rain, groundwater, and snowmelt runoff during the year with two dominant annual flow peaks corresponding to summer rainfall and early spring snowmelt. The average annual discharge of the Vistula below the mouth of the last large Carpathian tributary (the San River) reaches 420 m^3 s^{-1}, while the maximum observed discharge is 7700 m^3 s^{-1}. The suspended-sediment load constitutes about 65–70% of the total sediment load of the river, and is supplied mainly by the Carpathian tributaries (Figure 11.1D, Table 11.1).

TABLE 11.1 Contribution of major morphological units to suspended load supply to the piedmont Vistula River. A – situation without mechanical pollution of rivers draining the mining areas and without dam reservoirs in the catchment, B – the real situation. For location of the gauging stations see Figure 11.1B

Reach of the river		Major morphological units in the piedmont Vistula catchment (%)		
		Carpathians	Lowland fore-Carpathians	Middle Poland Upland
To the Zawichost gauging station	A	98	0	2
	B	89	1	10
To the Pulawy gauging station	A	96	0	4
	B	88	1	11

Human impact on fluvial processes in the piedmont Vistula valley can be divided into the following main phases:

1. The first and longest was the phase of systematically increased sediment load and sedimentation rates due to agricultural expansion in the region. This phase began in the Bronze Age (Gebica and Starkel, 1987; Klimek, 1988) and continued until the second half of the 19th century when channelization of the river was undertaken. The meandering course of the river was changed into a less sinusoidal or braided one, and the channel became much shallower and wider.
2. The second phase of the piedmont Vistula channel remodelling started after the 1840s due to construction of river regulation works, the intensity of which increased after a catastrophic flood occurred in 1884 (Starkel, 1982). In order to control river flow, to prevent natural channel adjustment, and to obtain a necessary depth for navigation, a variety of engineering methods have been employed:

   - the shortening of the river by cutting off selected meanders (the 19th century);
   - the building of stone groynes, perpendicular to the banks, to produce river channel narrowing (since the 1840s);
   - the construction of flood control embankments to limit the area of flooding (since the 1890s); and
   - the building of dams on the river and its tributaries for river flow regulation, water retention, flood protection and power generation (since the 1920s on the tributaries, and since the 1950s on the Vistula itself).

The most intense river channelization occurred between the 1890s and the 1960s. In the years 1920–1960 the channel materials were extensively exploited (Starkel, 1982; Klimek, 1987c). This phase of the channel remodelling is still continuing and has caused marked changes in the fluvial processes of the Vistula.

DATA USED

Recent trends in the changing geomorphology of the piedmont Vistula valley floor have been determined from large-scale maps showing the river channel changes over the last three centuries, from data based on surveys conducted during the river channelization in the 19th and 20th centuries (data obtained from the Polish Hydrological Survey), from large-scale morphological mapping of selected segments of the valley floor, from drilling of young *mada* deposits, and from data from the publications cited.

The following basic data have been analysed:

(i) Daily records of river stage since the beginning of measurements (gauging station in Cracow, 1813) till 1990 from all gauging stations on the river reach analysed (see Figure 11.1C).
(ii) Daily discharges during the period 1931–1990 from the gauging stations noted in (i).
(iii) Flow velocities measured (since 1931) and estimated (since about 1850) from selected gauging stations on the river.

(iv) Records of daily suspended load from all gauging stations on the piedmont Vistula and its tributaries, from 1946 until 1990.
(v) The results of repeated levelling surveys across the inter-embankment zone at each of the gauging stations during the period 1905–1990.

The large-scale maps analysed in the study are from the following years: 1657, 1737, 1782, 1785, 1804, 1810, 1815, 1817, 1824, 1839, 1855, 1866, 1882, 1900, 1933, 1953, 1975 and 1990 and most of them cover the entire segment of river discussed in this paper.

RESULTS

Figure 11.2 schematically explains changes in the fluvial processes of the piedmont Vistula, which have occurred during the period of human impact on the river.

Channel changes prior to the river channelization

Due to agricultural expansion in the river catchment (in the Carpathians especially), the frequency and rates of floods, channel aggradation and overbank sedimentation, channel shallowing and widening, and the size of meanders, systematically increased until the Little Ice Age (Gebica and Starkel, 1987; Kalicki and Starkel, 1987; Sokolowski, 1987). Until the end of the Middle Ages the meandering course of the piedmont Vistula changed little. Later, the flood discharges of the river increased by as much as three times (Starkel, 1982). During the Little Ice Age the Vistula channel became much larger (Falkowski, 1975, 1982; Maruszczak, 1982; Sokolowski, 1987; Babinski and Klimek, 1990). Downstream of the first large Carpathian tributary confluence (the Dunajec River) the course of the piedmont Vistula straightened. Large-scale maps from the second half of the 18th century show the Vistula as a river with a well-formed sinusoidal channel pattern downstream of that tributary. Downstream of the last Carpathian tributary confluence (the San River) the Vistula channel was remodelled into a braided pattern in the Little Ice Age (Falkowski, 1975, 1982; Maruszczak, 1982). This stage of the river channel is documented by maps from the 18th and 19th centuries (Figure 11.2C). Data obtained from drilling in numerous infilled meander cutoffs of various ages, and also from an analysis of the aforementioned maps, provide evidence that these trends in the piedmont Vistula channel geomorphology reached an advanced stage by the 19th century but were arrested later by river channelization. During the Little Ice Age, wide silty-sandy levees covered the floodplain and locally changed the river network.

Data resulting from the author's drilling of numerous infilled meander cutoffs of various ages provide the basis for interpreting changes in the piedmont Vistula channel geometry which occurred between the Atlantic Period and the start of river channelization (Figure 11.3). Detailed characteristics of the geometry of infilled abandoned meander channels of the piedmont Vistula formed before the Little Ice Age (e.g. Maruszczak, 1982; Gebica and Starkel 1987; Kalicki and Starkel, 1987; Rutkowski, 1987; Sokolowski, 1987), have allowed an estimate of the probable ages of the abandoned meanders located along the entire course of the piedmont Vistula.

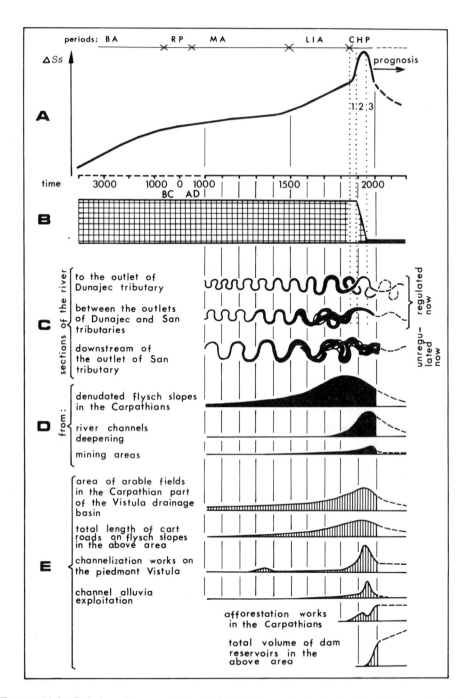

FIGURE 11.2 Relative changes of the sedimentation magnitude in the piedmont Vistula valley during the last millennium were initiated by human impact. These changes are: A – the relative changes of the sedimentation magnitude, B – the relative changes in the flooded area width, C – the changes of river channel pattern, D – the relative changes of sediment supply rates to the river, E – the relative changes of factors influencing this supply. Time periods: BA – Bronze Age, RP – Roman Period, MA – Middle Ages, LIA – Little Ice Age, CHP – channelization period of the river

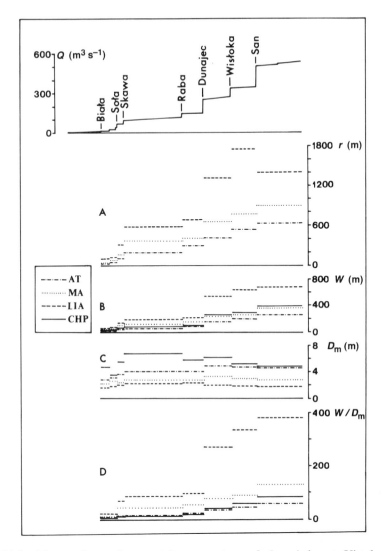

FIGURE 11.3 Mean values of geometric parameters of the piedmont Vistula channel estimated for successive river reaches located between mouths of the main Carpathian tributaries, during periods: AT – Atlantic Period, MA – Middle Ages, LIA – Little Ice Age, CHP – channelization period. Geometric parameters of the channel: r – radius of meanders, W – channel width, D_m – channel mean depth, W/D_m – compactness of the channel. Q – mean discharge from the years 1931–1990

The ages of meanders cut off since the 18th century, however, are more reliably estimated from the maps cited; these were produced every few to a dozen or so years.

Completely infilled meanders with small radius, which neighbour the whole river course, were formed before the Little Ice Age, and most of them are typical of the Atlantic Period. Others with larger radius have been recognized as meanders formed during the Middle Ages. Still larger meanders formed during the Little Ice Age are also completely infilled, and occur in numerous places along the study reach.

Prior to channelization of the river, the piedmont Vistula channel mean depth halved, on average, since the Atlantic Period (Figure 11.3C). Simultaneously, the channel mean width increased by a factor of three, on average (Figure 11.3B). At the same time, the width/depth ratio (form ratio) increased sixfold and reached a value of 380 (Figure 11.3D). The average radius of the large meanders formed downstream of the largest Carpathian tributaries during the Little Ice Age reached 1200–1800 m, and exceeded its average value from the Atlantic Period by three times (Figure 11.3A).

The maps from the years 1780–1850 provide a basis for estimating the *channel-forming discharge*, Q_{chf} of the pre-regulated piedmont Vistula. Among numerous parameters of the river channel which represent its state prior to river channelization, only map-documented meander wavelength is known precisely. For this reason Inglis's (1941) formula ($L = 29.6\sqrt{Q_{chf}}$, where L = meander wavelength and Q_{chf} = channel-forming discharge) has been used for the necessary calculations; results are presented in Table 11.2 An analysis of *bankfull discharge*, Q_{bf}, the calculation of which is explained elsewhere (Lajczak, 1994c, 1994d), shows its values essentially are equal to the mean high discharge, MHQ, of the pre-regulated river. Only in the presently meandering river reach (upstream of the first large Carpathian tributary, the Dunajec River) are Q_{chf} and Q_{bf} approximately equal. Downstream, in the straightened reach, and even in the braided channel reaches with relatively large width, the channel was formed by discharges as much as two times greater than the bankfull discharge.

Accelerated sediment supply to the Vistula channel from deforested, cultivated and strongly denuded Carpathian flysch slopes, which occurred after the 17th century (Falkowski, 1975, 1982; Adamczyk, 1978, 1981; Klimek, 1987c), was

TABLE 11.2 Calculated and estimated characteristic values of discharge Q (m^3 s^{-1}) of successive reaches of the piedmont Vistula for its pre-regulation stage. MQ – mean discharge, MHQ – mean high discharge, Q_{bf} – bankfull discharge, Q_{chf} – channel-forming discharge. For location of the tributaries to the Vistula mentioned above see Figure 11.1B

Reaches of the river between mouths of the following tributaries	Measured Q values from years 1931–1990			Estimated Q values before the river channelization	
	MQ	MHQ	MHQ/MQ	$Q_{bf} \approx MHQ$	Q_{chf}
Biala – Gostynka	13	39	3.1	39	26
Gostynka – Sola	23	60	2.7	60	48
Sola – Skawa	68	163	2.4	163	128
Skawa – Raba	104	215	2.1	215	186
Raba – Dunajec	138	287	2.1	287	254
Dunajec – Wisloka	262	517	2.0	517	930
Wisloka – San	340	656	1.9	656	1200
San – Pulawy gauging station	516	1000	1.9	1000	1450

associated with increasing size and frequency of floods on the piedmont Vistula floodplain. The 19th century was characterized by severe floods, which repeatedly affected the floodplain (Mikulski, 1954). As a result, fine material was deposited on the floodplain surface. In addition, lateral migration of the shallow and aggrading piedmont Vistula channel meanders reached relatively high rates from the 17th century to the 19th century. The results of the above calculations based on detailed analysis of successive large-scale maps from the 17th to 20th centuries, are presented in Table 11.3.

Channel and floodplain remodelling due to river channelization

The second phase of the river channel and floodplain remodelling started immediately after river channelization was introduced. This very short phase distinguishes itself by the remarkably altered hydraulic conditions of the river and the rapid changes in the form of the channel and the floodplain. After completion of the flood control embankments, the inter-embankment zone acted as an active floodplain, albeit much narrower than the older flooded area. A new channel equilibrium is being established in some reaches of the river.

Response of the river to channel regulation

The elimination of numerous meanders shortened the river in the study reach by 15%, and locally (upstream of the mouth of the Sola tributary) by as much as 75%. Shortened study reaches of the piedmont Vistula due to the channelization are shown in Figures 11.4 and 11.5 and summarized results of the river channelization are shown in Figure 11.6. The channel slope clearly increased, and in numerous reaches has stabilized at twice its former value.

A steeper channel, completed groyne system, and the extensive extraction of channel sediment, together initiated rapid channel deepening. Channel narrowing has followed a complete silting of the inter-groyne basins (Figure 11.7) and older banks subsequently have accumulated sandy levees. A silty-sandy sediment load, originating partially from the Vistula channel downcutting, has fossilized an older

TABLE 11.3 Mean values of concave bank removal (m yr^{-1}) of selected meanders of the piedmont Vistula in separate periods representing the pre- and post-regulation stages of the river

Reaches of the river between mouths of the following tributaries	Period			
	1657–1780	1780–1840	1840–1900	1900–1992
Sola – Skawa	–	3.23	0.83	0.58
Skawa – Raba	(3.23)?	3.57	1.52	0.70
Raba – Dunajec	–	3.97	1.32	0.41
Dunajec – Wisloka	–	5.55	2.45	0.85
Wisloka – San	–	7.08	2.40	1.23

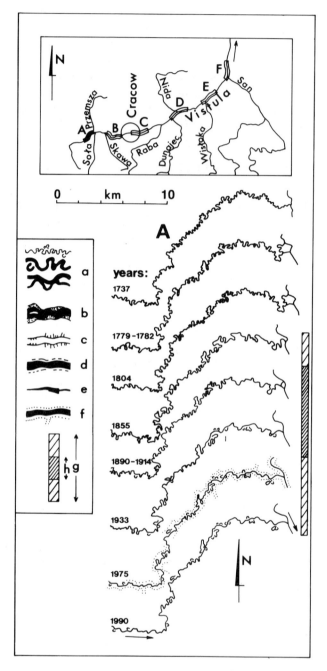

FIGURE 11.4 Changes in the channel pattern of the piedmont Vistula within the Oświęcim Basin due to the river channelization. Note the significant shortening of the channel length which occurred after 1900. a, b – river channel, c, groyne system, d – wet places behind the levees, e – dam reservoirs, f – flood control embankments, g – period of river channelization, h – period of the most intensive river regulation works

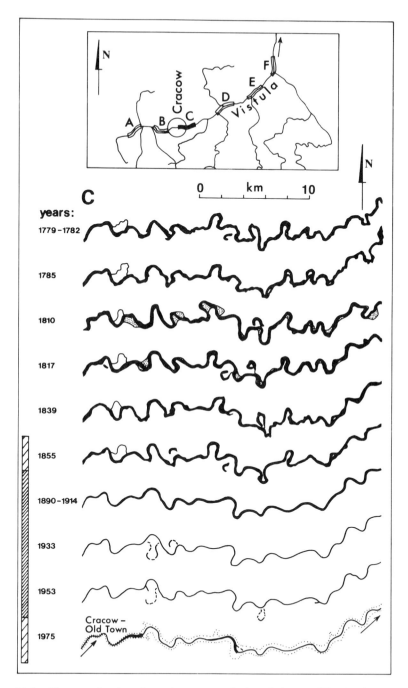

FIGURE 11.5 Changes in the channel pattern of the piedmont Vistula downstream of Cracow due to the river channelization. The figure shows the channel shortening and narrowing started from the middle of the 19th century. See Figure 11.4 for the symbol key

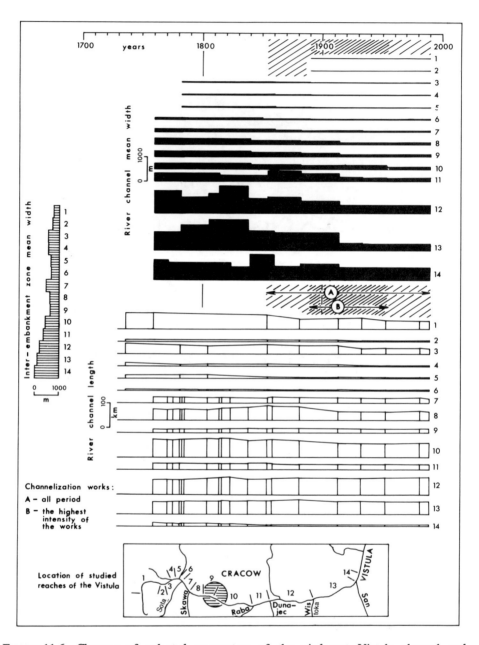

FIGURE 11.6 Changes of selected parameters of the piedmont Vistula channel and floodplain over the last 250 years were estimated on the basis of the large-scale maps analysed. A decreasing trend in channel length and mean width of separate river reaches was initiated about 1900 due to the river channelization. A – whole period of the river channelization, B – period of the most intensive river regulation works. The bottom figure shows the location of the analysed river reaches

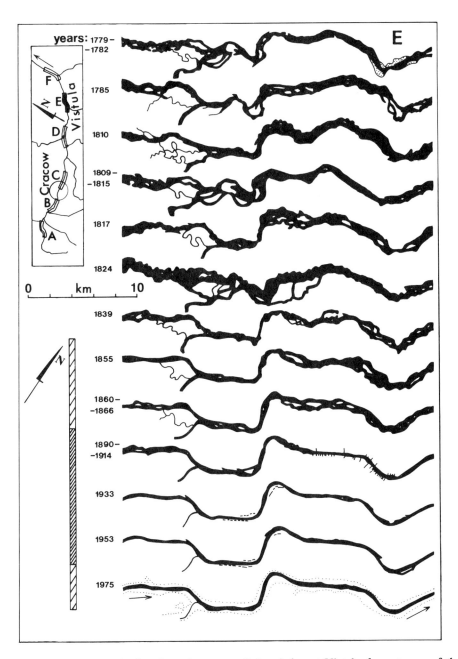

FIGURE 11.7 Changes in the channel pattern of the piedmont Vistula downstream of the mouth of the Wisloka River due to river channelization. Note the significant narrowing of the channel due to the inter-groyne basin sedimentation which occurred after 1860. For symbol key, see Figure 11.4

floodplain morphology within the inter-embankment zone. This process, started after 1890, is illustrated by Figure 11.8, which shows the deepening and narrowing of the piedmont Vistula channel.

The rate of post-channelization deepening of the channel is variable along the course of the river (Figure 11.9). These rates vary in concert with the mean thickness of overbank sedimentation (including the inter-groyne basin filling). The piedmont Vistula is characterized by relatively high rates of channel deepening, reaching almost 4.5 m near Cracow, and also a relatively high mean thickness of overbank sedimentation (up to 4 m). These rates are two or three times greater in relation to those recorded by Babinski (1984, 1992) in the regulated Lower Vistula.

The present-day deep river channel, with high and steep banks, is characterized by slow lateral migration (Table 11.3). Incised meanders formed earlier have been stabilized and this process is aided by bank protection works. The stabilized horizontal position of the river channel has favoured the levee growth.

The piedmont Vistula channelization has reversed the earlier trend in the morphological development of the river channel. The mean width of the remodelled river channel currently approximates the values typical of the channel formed before the Little Ice Age (Figure 11.3). The mean depth of the channel just exceeds the channel depths typical of the Atlantic Period, and in the most overdeepened reach the river is almost twice as deep. Finally, the present-day shape of the channel, expressed by the W/D_m ratio, resembles that of the channel from the Atlantic Period, especially the most overdeepened reach of the channel with its typical "box" cross-section.

Remodelling of the piedmont Vistula channel has increased flow velocity of the river. Over the last 150 years the flow velocity has as much as doubled in the case of mean discharges, and as much as tripled in the case of high discharges (Lajczak, 1994c, 1994d). The trend of increasing flow velocity with respect to time is still continuing due to changes of the channel geometry (increases in D_m and decreases in W/D_m; Lajczak, 1991).

Increasing flow velocity within the overdeepened channel, in spite of a decrease in cross-sectional area of the channel, has increased bankfull discharge. Cross-section surveys of the channel at gauging stations, repeated every few years, have allowed an estimate of the change of the bankfull stage with respect to time. Using successive rating curves for the gauging stations (altered by flood scouring) the Q_{bf} values (Q related to bankfull stage) can easily be calculated. The river channelization on the piedmont Vistula has increased bankfull discharge threefold, on average (Lajczak, 1994c, 1994d). The pattern of increased Q_{bf} is presented in Figure 11.10 for selected gauging stations.

FIGURE 11.8 The piedmont Vistula channel deepening and narrowing as exemplified by changes in channel geometry measured in selected gauging stations: 1 – Goczalkowice, 2 – Tyniec, 3 – Sandomierz. The decrease of the minimum annual water level WL_{min} (A) and the cross-profile lowering (B), indicate that this deepening was caused by the river channelization. C – the relationships between channel level changes were calculated on the basis of WL_{min} values (ΔH_1) and the mean levels of channel determined from levelling (ΔH_2). Channel level changes between successive level measurements are shown by large points in the diagram. L – left bank, R – right bank

River Regulation and Response of the Upper Vistula River

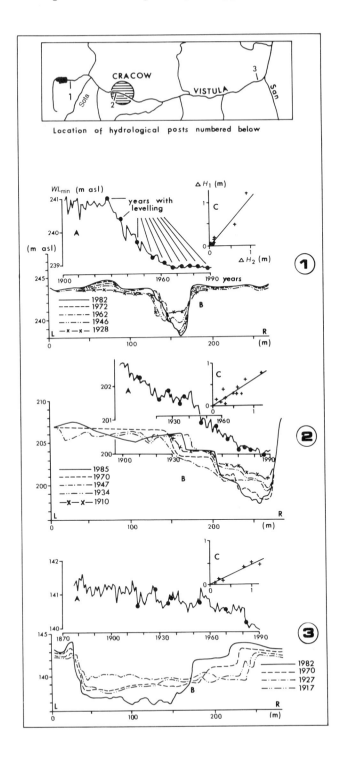

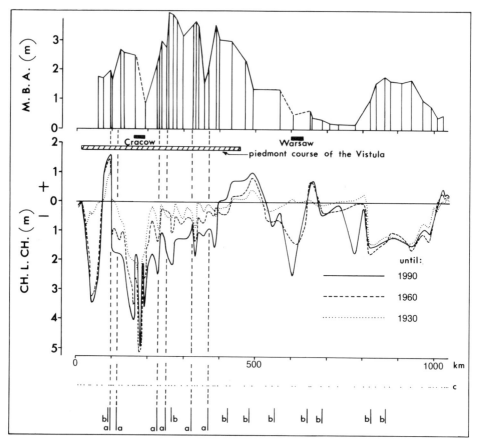

FIGURE 11.9 Rates of channel deepening initiated by river channelization, represented by minus values of the channel level changes (CH.L.CH.), and the mean thickness of overbank sedimentation, represented by mean rates of inter-groyne basin sedimentation and levee growth (M.B.A.), which have occurred along the whole course of the Vistula. The rates discussed are from repeated channel cross-section levelling, a – mouths of main Carpathian tributaries, b – mouths of main upland and lowland tributaries, c – location of analysed gauging profiles on the Vistula

Before river channelization, bankfull discharge of the piedmont Vistula was approximately equal to its mean high discharge, *MHQ*, calculated for the long period after introduction of channelization (Figure 11.10). Bankfull discharge subsequently increased to equal channel-forming discharge in the meandering reach of the river, while Q_{bf} was almost half that of Q_{chf} in the straightened segment of the river (Table 11.2). Now, the remodelled piedmont Vistula channel is formed by discharges three times greater, on average, and the increased trend in bankfull discharge continues. This trend reflects continuous channel deepening (Figure 11.10).

Currently, much larger amounts of suspended load and bedload, originating from the continuous channel deepening, are transported over longer distances due to increasing river competence. A study of long-term changes in suspended-sediment

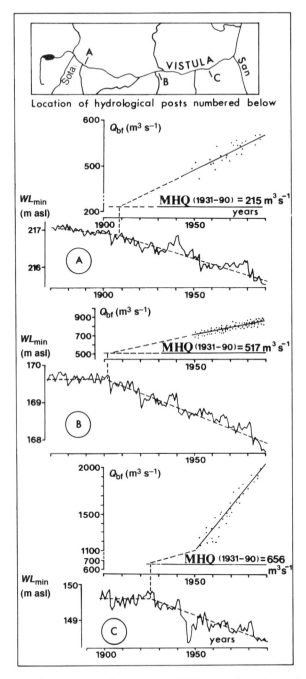

FIGURE 11.10 The minimum annual water level WL_{min} at selected gauging stations on the piedmont Vistula indicating the lowering of the channel. The gauging stations are: A – Smolice, B – Karsy, C – Kolo. Channel deepening causes continuous increase in bankfull discharge, Q_{bf}, compared with a mean high discharge, MHQ

transport of the Vistula provides evidence that the largest amounts of that material were transported during the first decades after the completion of river confinement (Lajczak, 1994c). At this time overbank sedimentation (including the inter-groyne basin filling) also reached very high rates (see Figure 11.2A).

Further human impact has favoured a decreasing trend in sediment transport rates, a trend that began about 1930. There are numerous reasons for this trend:

(i) advantageous changes in land use in the catchment basin (an increase in the area of pastures and meadows and, especially, reforestation projects in the Carpathians);
(ii) significant lowering of the downcutting rates in some reaches of the river channel within the Oświeçim Basin, started after the 1950s (see Figure 11.8);
(iii) the increasing number of dams on the Vistula and its tributaries, especially the Carpathian rivers, trapping large amounts of the sediment load in the reservoirs.

Due to the deep channel and more rapid hydrologic response, shortening of the length of the inter-embankment floodplain inundation period has been recorded (Figure 11.11). Now the active piedmont Vistula valley floor is flooded only in summer. Before river channelization the floodplain was inundated for much longer and the total period of flooding during snowmelt floods and summer floods was similar. Consequently there is a decrease in overbank sedimentation rates within the inter-embankment floodplain and a decrease in rates of levee growth (Figure 11.11).

Channel and floodplain remodelling

The present-day development of the morphologically active piedmont Vistula floodplain occurs in a manner different from the situation characterizing the river before channelization was undertaken. The development of floodplain morphology over the last 150 years, since the introduction of river channelization, can be divided into three subphases (see Figure 11.2):

1. *1840s–1890s:* The elimination of numerous meanders upstream of the Dunajec confluence and the groyne building in selected segments of the river banks, was initiated. This started local channel deepening associated with rapid sedimentation in the inter-groyne basins. The channel deepening of the piedmont Vistula started locally earlier in the 18th century due to the construction of cutoffs on selected meanders. Local intensive channel regulation works on the short Vistula channel reach in Cracow were introduced after the catastrophic summer flood in 1813 and they initiated a continuous channel deepening recorded in the city from 1820. In the 1840–1890s the direction of the previous piedmont Vistula channel development was locally changed. During each flood event the wide valley floor was inundated and covered by fine material.

2. *1890s–1950s*: During this period the channel and floodplain were changed along the entire course of the piedmont Vistula. Continuation of groyne and flood control embankment construction along almost all of the river reach analysed, as well as the continued cutting off of meanders (Figures 11.4 and 11.5), produced a rapid channel deepening and narrowing associated with simultaneous inter-groyne basin filling and

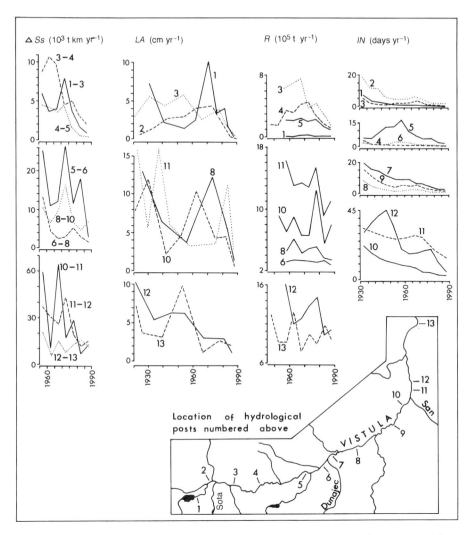

FIGURE 11.11 Decreasing rates of suspended-load transport, R, and sedimentation, ΔSs, the time of the inter-embankment floodplain inundation, IN, and the levee growth, LA, determined from successive gauging station and floodplain surveys on reaches of the piedmont Vistula for the last seven decades. The decreasing trend results from continuous channel deepening

old bank fossilization (Figure 11.8). In cases where bankfull discharge and flow velocity increased, a vast volume of sandy material was deposited along the banks. This material built steep levees and plugs. The flooded area, subject to much more intensive overbank sedimentation, was reduced to a very narrow inter-embankment zone. After completion of the flood control embankments the area located outside the embankments was not significantly built up. Meander cutoffs and older oxbow lakes were completely infilled by silt-clay deposits (Figure 11.12).

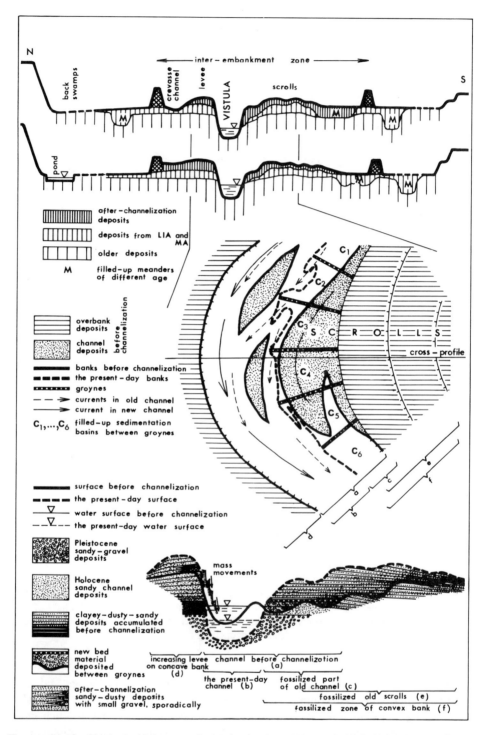

FIGURE 11.12 Typical thickness of overbank deposits accumulated within the interembankment floodplain of the piedmont Vistula over the last 100 years. The remodelled channel and floodplain due to river channelization, and changes in the character of sediments deposited before channelization and during the last century, are shown

Permanent channel deepening and levee growth increased local relief and unevenness on the floodplain surface. Mass movements have become more active on the steepened banks, particularly above the contact between deposits of channel facies and overbank facies, which now stands high above the water surface of the river. Drilling has shown that the steep inclinations of levee slopes, when compared with those in the pre-regulation stage, have resulted from the deposition of increasingly coarse material on the floodplain. Furthermore, the increased role of the summer flood in the total inundation period of the floodplain must be recognized as the second reason for the steepening of the levees. The higher roughness of rich summer vegetation compared with that related to early spring plants more effectively limits the distance of coarse-particle transportation on the floodplain surface. Steepening of levees by crevasse channels, which are now new landforms located marginally within the inter-embankment floodplain with a limited width of overbank flow, is the third reason for the increase in the size of the levees (Figure 11.12). These factors in concert have increased the relative heights of levees by as much as 3–4 m, and slopes by 10–20°, and even as much as 40° locally. If the wide sandy levees migrate to the flood control embankments, the inter-embankment floodplain level may be as high as 2 m above the surface outside the embankments.

3. *After the 1950s*: During the last four decades, and in the case of the reach of channel near the Upper Silesia – Cracow mining region, after the 1970s, a significant trend of declining suspended-sediment load has been recorded (Lajczak, 1994c). Only in the most overdeepened reaches of the piedmont Vistula channel, incised to the gravel bed material, has the previously rapid rate of change in channel and floodplain morphology declined. This suggests that these deepest reaches of the channel have established a new equilibrium involving higher and stabilized channel slope and very slow lateral channel migration. The present-day morphology of the active valley floor allows for decreased rates of overbank sedimentation because the drainage works within the flooded area remove sediment-laden overbank flows. Local depressions, active crevasse channels and silted oxbow lakes within the inter-embankment zone are drained of muddy water via drainage ditches to the deepened river channel only a few days after the peak flows.

CONCLUSIONS

The evolution of the valley-floor landscape of the piedmont Vistula during the period analysed in this study is similar to the geomorphic history of other river valley floors and alluvial plains (Wolman, 1967; Trimble, 1976; Walling, 1974; Gregory, 1987; Knox, 1987; Dai, 1988). Indeed, the evolution of the Vistula is quite comparable to the generalized evolution of the Appalachian Piedmont, USA, 1700–1970, described by Trimble (1970).

The most rapid changes in the relief of the piedmont Vistula valley floor occurred during the first few years after the river was confined; the river channelization reversed a previous trend in the channel development. Channel deepening and narrowing developed after numerous meanders were cut off and the groyne system had been initiated. Channel downcutting is still continuing. Simultaneously, the channel downcutting accelerated sedimentation in the inter-embankment floodplain zone. River channelization is acknowledged as the most significant cause of the changes in the river sediment load and sedimentation rates, which have occurred in the Vistula during the period of human impact. The cascade reservoir system on the river and its tributaries, especially the Carpathian rivers, was developed a few decades ago and markedly reduced the rate of sediment transport. This suggests that river channelization causes significant acceleration of channel erosion and overbank sedimentation, especially during the first years after initiation of the regulation works; later the changes slow (Lajczak, 1994c). The demonstrated inter-dependence between sediment load and sedimentation rate in the Vistula River, confirms the role of river erosion in influencing overbank deposition rates (McHenry, 1974). The causes of rapid alluviation of the floodplain noted in this study compares with other examples of intensive urbanization in a river catchment, such as those reported by Wolman (1967), Walling (1974), Trimble (1976) and Gregory (1987).

The response of the piedmont Vistula channel to river channelization is similar to that on the Lower Vistula channel reported by Babinski (1984, 1985, 1992). Channel deepening and overbank sedimentation of the piedmont Vistula reached two or three times corresponding rates on the Lower Vistula during the channelization period. The inter-groyne basins of the Lower Vistula are infilled by bed material, while the growth of new surfaces within the inter-embankment floodplain of the piedmont Vistula is caused by increased deposition of fine-grained material, a situation typical of areas located adjacent to strongly denuded mountains (Lajczak, 1994d).

If decreasing rates of channel downcutting and overbank sedimentation affect additional segments of the piedmont Vistula valley floor, slow channel widening due to mass movements on steep banks will stabilize the channel as a new equilibrium is reached. In the long-term, however, very slow channel shoaling may subsequently be induced by the much reduced lateral migration rate of the channel.

The morphological development of the Vistula valley floor described here will be modified again by the construction of new dams on the river and its tributaries. A completed system of planned reservoirs will reduce sediment input to the Vistula channel by as much as 100 times (Lajczak, 1993). Previous response to the development of six shallow reservoirs on the Vistula near Cracow suggests that the piedmont Vistula channel with a full cascade system of shallow reservoirs will slowly aggrade, especially after a few decades of operation (Lajczak, 1994b). If deep reservoirs are completed on the lower reaches of the Carpathian tributaries to the Vistula, further siltation of reservoirs on the Vistula will reach extremely low rates. The cascaded Vistula channel will thus reach a more stable vertical position. In addition, stabilized banks will stop the lateral migration of the channel.

Thus the completion of a full cascade system on the piedmont Vistula will initiate yet another phase of channel adjustments to human impact.

ACKNOWLEDGEMENTS

The study has been supported by the grant 2.3. of the Polish Academy of Sciences. Professor Kazimierz Klimek, Polish Academy of Sciences in Cracow, is gratefully acknowledged for his stimulating discussions and valuable advice, and for helpful criticism of the manuscript. I would like to extend my deepest gratitude to Professor Ake Sundborg, Uppsala University, for his helpful comments and suggestions concerning the problems discussed in the study. Thanks are also due to two anonymous referees for their critical comments on the manuscript. Mrs A. Scott-Brown, of the International House in Cracow, is very kindly acknowledged for improvement of the English version of the paper. I wish to thank the Polish Hydrological Survey for providing the primary measurement data. I am also grateful to my son, Leszek, who kindly assisted in the field.

REFERENCES

Adamczyk, M. J. 1978. Changes in landscape of the Polish Carpathians in 1650–1870. *Wierchy*, **47**, 160–176 (in Polish).

Adamczyk, M. J. 1981. Rivers of the North Carpathians in the 18th and 19th centuries. *Wierchy*, **49**, 227–240 (in Polish).

Babiński, Z. 1984. The effects of human activity on changes in the Lower Vistula channel. *Geogr. Pol.*, **50**, 271–282.

Babiński, Z. 1985. Hydromorphological consequences of the Lower Vistula regulation. *Przegl. Geogr.*, **57**, 471–486 (English summary).

Babiński Z. 1992. The present-day fluvial processes of the Lower Vistula River. *Geogr. Studies*, **157**, 1–171 (English summary).

Babiński, Z. and Klimek, K. 1990. The present-day channel and floodplain of the River Vistula. In: Starkel, L. (ed.), Evolution of the Vistula River valley during the last 15 000 years. *Geogr. Studies Spec. Issue* No. 5, pp. 62–75.

Brookes, A. 1990. *Channelized rivers: Prospectives for Environmental Management.* Wiley, Chichester, 326 pp.

Castiglioni, G. B. and Pellegrini, G. B. 1981. Two maps on the dynamics of a river bed. *Proceedings of International Symposium on Erosion and Sediment Transport Measurement*, Florence.

Dai, S. 1988. Analysis of sediment yields during the historic period in the loess region of the Yellow River basin. In: *Sediment Budgets*, IAHS Publ. 174, pp. 377–380.

Falkowski, E. 1975. Variability of channel process of lowland rivers in Poland and changes of the valley floors during the Holocene. *Biul. Geol.* UW, **19**, 45–78.

Falkowski, E. 1982. The pattern of changes in the Middle Vistula valley floor. In: Starkel, L. (ed.), Evolution of the Vistula River valley during the last 15 000 years. *Geogr. Studies, Spec. Issue* No. 1, pp. 79–92.

Gębica, P. and Starkel, L. 1987. Evolution of the Vistula River valley at the northern margin of the Niepolomice Forest during the last 15 000 years. In: Starkel, L. (ed.), Evolution of the Vistula River valley during the last 15 000 years. *Geogr. Studies, Spec. Issue* No. 4, pp. 71–86.

Gong, S. 1987. The role of reservoirs and silt-trap dams in reducing sediment delivery into the Yellow River. *Geogr. Ann.*, **69A**, 173–179.

Graf, W. L. 1975. The impact of suburbanization on fluvial geomorphology. *Water Resources Research*, **11**(5), 690–692.

Gregory, K. J. 1987. River channels. In: Gregory, K. J. and Walling, D. E. (eds), *Human Activity and Environmental Process*, Wiley, Chichester, pp. 207–235.

Gregory, K. J. and Brookes, A. 1983. Hydromorphology downstream from bridges. *Appl. Georg.*, **3**, 145–159.

Gregory, K. J. and Madew, J. R. 1982. Land use change frequency and channel adjustment. In: Hey, R. D., Bathurst, J. C. and Thorne, C. R. (eds), *Gravel-Bed Rivers*, Wiley, Chichester, pp. 757–781.

Hammer, T. R. 1972. Stream channel enlargement due to urbanization. *Water Resources Research*, **8**(6), 1530–1540.

Hickin, E. J. 1983. River channel changes: Retrospect and prospect. In: Collinson, J. D. and Lewin, J. (eds), Modern and ancient fluvial systems. *Spec. Pub. Intern. Ass. Sediment*, **6**, 61–83.

Inglis, C. C. 1941. Meandering of rivers. *Central Board of Irrigation, India*, 24.

Kalicki, T. and Starkel, L. 1987. The evolution of the Vistula River valley downstream of Cracow during the last 15 000 years. In: Starkel, L. (ed.), Evolution of the Vistula River valley during the last 15 000 years. *Geogr. Studies Spec. Issue* No. 4, pp. 51–70.

Klimek, K. 1987a. Holocene history of the Vistula valley within the Oświęcim Basin, Carpathians Foreland, Poland. *Proc. of IGCP 158 Symp., Sweden, Abstr. of lect. and post.*

Klimek, K. 1987b. Vistula valley in the eastern part of the Oświęcim Basin during the Upper Vistulian and Holocene. In: Starkel, L. (ed.), Evolution of the Vistula River valley during the last 15 000 years. *Geogr. Studies Spec. Issue* No. 4, 13–29.

Klimek, K. 1987c. Man's impact on fluvial processes in the Polish Western Carpathians. *Geogr. Ann.*, **69A**(1), 221–226.

Klimek, K. 1988. An early anthropogenic alluviation in the subcarpathian Oświęcim Basin, Poland. *Bull. of the P.A. of Sci., Earth Sciences*, **36**(2), 159–169.

Knighton, A. D. 1989. River adjustment to changes in sediment load: the effects of tin mining on the Ringarooma river, Tasmania, 1875–1984. *Earth Surface Processes and Landforms*, **14**, 333–359.

Knighton, A. D. 1991. Channel bed adjustment along mine-affected rivers of northeast Tasmania. *Geomorphology*, **4**, 205–219.

Knox, J. C. 1987. Historical valley floor sedimentation in the upper Mississippi valley. *Annals of the Association of American Geographers*, **77**, 224–244.

Lajczak, A. 1991. Flow velocity rates in longitudinal profile of the Vistula River. *Wiad. Inst. Meteor. i Gosp. Wodn.*, **14**, 55–67 (in Polish).

Lajczak, A. 1993. Planning of the multi-function dam reservoirs distribution in the large river basin: criticism of the present-day stage and proposition for reduction the silting process. The Vistula basin, Poland. *Proc. of the Third Intern. Conf. and Exhib. on Small Hydropower*, **4**, 69–80, Munich.

Lajczak, A. 1994a. The rates of silting and the useful lifetime of dam reservoirs in the Polish Carpathians. *Zeitschrift für Geomorphologie*, **37**(2), 129–150.

Lajczak, A. 1994b. Modelling the long-term course of reservoir siltation and estimating the life of dams. *Earth Surface Processes and Landforms* (submitted).

Lajczak, A. 1994c. Man's induced changes of river load transportation and sedimentation rates by the Vistula river, Poland *Annals of the University Maria Curie-Sklodowska*, Lublin, (in print).

Lajczak, A. 1994d. Potential rates of the present-day overbank sedimentation in the Vistula valley at the Carpathian foreland, Southern Poland. *Quaest. Geogr.* (in print).

Maruszczak, H. 1982. The Middle Vistula near Lublin. In: Piskozub, A. (ed.), *The Monograph of the Vistula River*, Warsaw, pp. 125–136, (in Polish).

McHenry, I. R. 1974. Reservoir sedimentation. *Water Res. Bull., Am. Water Res. Ass.*, **10**(2), 329–337.

Mikulski, Z. 1954. Catastrophic floods in Poland. *Czas. Geogr.* **25**, 380–396 (English summary).

Park, C. C. 1977. Man-induced changes in stream channel capacity. In: Gregory, K. J. (ed.), *River Channel Changes*, Wiley, Chichester, pp. 121–144.

Rutkowski, J. 1987. Vistula river valley in the Cracow Gate during the Holocene. In: Starkel, L. (ed.), Evolution of the Vistula River valley during the last 15 000 years. *Geogr. Studies, Spec. Issue* No 4, pp. 31–49.

Schumm, S. A. 1968. River adjustment to altered hydrologic regimen – Murrumbidgee River and paleochannels, Australia. *US Geological Survey Professional Paper*, **598**, 1–65.

Sokolowski, T. 1987. Vistula valley between the outlets of Dunajec and Breń rivers. In: Starkel, L. (ed.), Evolution of the Vistula River valley during the last 15 000 years, *Geogr. Studies, Spec. Issue* No. 4, pp. 95–114.

Starkel, L. 1982. The Upper Vistula downstream of Cracow. In: Piskozub, A. (ed.), *The Monograph of the Vistula River*, Warsaw, pp. 111–124, (in Polish).

Starkel, L. 1983. The reflection of hydrologic changes in the fluvial environment of the temperate zone during the last 15 000 years. In: Gregory, K. J. (ed.), *Background to Palaeohydrology*. Wiley, Chichester, pp. 213–235.

Starkel, L. 1987. Anthropogenic sedimentological changes in Central Europe. *Striae*, **26**, 21–29.
Strahler, A. N. 1956. The nature of induced erosion and aggradation. In: Thomas, W. L. (ed.), *Man's role in Changing the Face of the Earth*, University of Chicago Press, pp. 621–638.
Sundborg, Å. 1983. Sedimentation problems in river basins. *Nature and Resources*, **19**(2), 10–21.
Sundborg, Å. 1986. Sedimentation processes. *Proc. of Intern. Symp. on Erosion and Sediment.* in Arab Countries, Baghdad, Iraq, pp. 1–27.
Szumański, A. and Starkel, L. 1990. Channel pattern changes and attempts to reconstruct the hydrological changes. In: Starkel, L. (ed.), Evolution of the Vistula River valley during the last 15 000 years, *Geogr. Studies, Spec. Issue* No. 5, pp. 154–163.
Trimble, S. W. 1970. The Alcovey River swamps. The result of culturally accelerated sedimentation. *Bull. Georgia Academy of Sci.*, **28**, 131–141.
Trimble, S. W. 1976. Modern stream and valley sedimentation in the Driftless area, Wisconsin, USA. *Intern. Geography' 76*, **1**, 228–231.
Walling, D. E. 1974. Suspended sediment and solute yields from a small catchment prior to urbanization. *Inst. of British Geogr. Spec. Publ.*, **6**, 169–192.
Wolman, M. G. 1967. A cycle of sedimentation and erosion in urban river channels. *Geogr. Ann.* **48A** (2–4), 385–395.
Wyżga, B. 1993. River response to channel regulation: Case study of the Raba River, Carpathians, Poland. *Earth Surface Processes and Landforms*, **18**, 541–556.

12

Morphological Changes in a Large Braided Sand-Bed River

E. MOSSELMAN

Delft Hydraulics, Emmeloord, The Netherlands

M. HUISINK, E. KOOMEN*

Institute for Earth Sciences, Free University, Amsterdam, The Netherlands

AND

A. C. SEIJMONSBERGEN

Landscape and Environmental Research Group, University of Amsterdam, The Netherlands

ABSTRACT

Processes governing morphological changes in the braided Brahmaputra–Jamuna River have been studied by using satellite images, discharge records and cross-sectional data. The correlation between annual erosion and magnitude and duration of the annual flood was weak. Channel abandonment was clearly correlated with upstream bifurcation geometry, but channel width adjustment was not. From this we conclude that channels are abandoned through shallowing rather than narrowing. Bend migration of individual channels could be simulated with a simple meander model. There are indications that channel patterns within the braid belt are influenced by tectonics.

INTRODUCTION

The Brahmaputra–Jamuna is a large braided sand-bed river that flows through eastern India and Bangladesh, see Figure 12.1. The width of its braid belt varies between 5 and 17 km and individual channels are up to 2 km wide. The annual flood is about 60 000 m^3 s^{-1}. Scour depths can reach 40 m and bank erosion rates can locally be as high as 1000 m a year. The associated rapid planform changes are very complex with bend migration, formation and propagation of bars, channel creation, channel abandonment, and migration of confluences and bifurcations. Previous studies of these

---

*Current address: Kastelenstraat 243 3h, 1082 EG Amsterdam, The Netherlands

River Geomorphology. Edited by Edward J. Hickin
© 1995 John Wiley & Sons Ltd

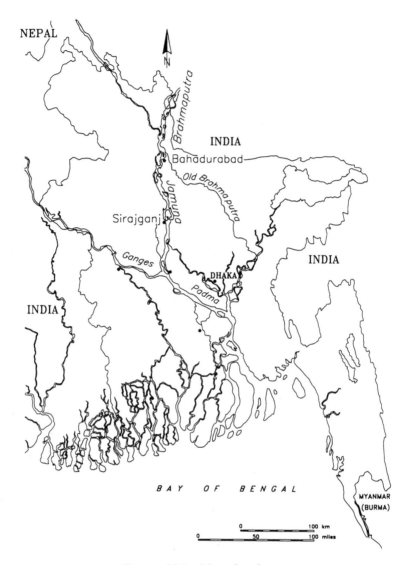

FIGURE 12.1 Map of study area

processes are reported by Coleman (1969), Bristow (1987), Klaassen and Vermeer (1988a, 1988b) and Klaassen and Masselink (1992). General considerations on predictability and model concepts are discussed by Klaassen et al. (1993).

Predictions of planform changes over a period of 1 to 5 years are needed for the construction of the Jamuna Multipurpose Bridge, the site selection for river training and bank protection works (Food Action Plan, FAP21) and active floodplain management by recurrent measures (FAP22). A probabilistic predictive model is currently being developed. It is based on submodels for constituent mechanisms on a high level of aggregation instead of submodels for fundamental laws of physics for small volumes of water and sediment. The aggregated mechanisms include width adjustment, channel abandonment, bend migration and the influence of tectonics. Here we present the main results from a preliminary study of these mechanisms. The study was carried out in 1992, but had to remain without follow-up in 1993 and 1994. We consider the present results as preliminary for two reasons. First, we intend to automate the analyses which would allow the inclusion of many more data. Second, a final conclusion on the performance and the relative importance of each of the constituent mechanisms can only be drawn after having combined the various submodels into one integrated model.

METHODS

General

For the study we use daily discharges from Bahadurabad station, cross-sections from the Bangladesh Water Development Board and an extensive set of satellite images from Landsat MSS, Landsat TM and MOS-MESSR sensors of the Brahmaputra–Jamuna River in Bangladesh. The images cover almost every year from 1973 to 1992. They were processed by the National Aerospace Laboratory NLR. This processing included the production of multitemporal change-detection images. We measure geometrical parameters of the channel network by hand from hard-copies of the images, but we intend to write computer programs for an automated determination of these parameters from the digital satellite data in the future. Such programs would make it feasible to include many more data.

Relation between bank erosion and discharge

The discharge through a particular channel within the braid belt can be estimated from the total by assuming that the individual discharges are distributed in proportion to the conveyance of each channel. Cross-sectional data, however, are not available for every channel, and even when a cross-section is available, it is not necessarily representative because of the strong non-uniformity of the channels. Indeed, our preliminary attempts to correlate morphological changes with conveyances yielded poor results. We therefore try to find a correlation between an overall measure of planform change and a parameter characterizing magnitude and duration of the total annual flood. For the former we use the total area of land eroded annually along 14 major bends and for the latter we follow the parameterization of the discharge hydrograph which arises from the one-dimensional calculation of longitudinal river-bed profiles in dynamic equilibrium. This parameterization is explained below.

Here we assume in the parameterization a constant width for all stages below bankfull and an infinitely large width at higher discharges when vast areas of the country are flooded and the rating curve becomes almost horizontal. The morphological effects of the higher discharges thus hardly differ from the effects of the bankfull discharge. Hence we truncate the higher discharges at the bankfull value of 44 000 $m^3 s^{-1}$, so that a constant width can be used throughout.

Discharges and sediment transport rates may fluctuate in a dynamic equilibrium, but the average amount of sediment transported, S_m, remains constant (cf. Jansen et al., 1979):

$$S_m = \frac{1}{T} \int_0^T B s_s(t) \, dt = \text{constant} \tag{12.1}$$

in which T is a sufficiently long time interval, B is the river width, s_s is the sediment transport per unit width and t denotes time. Assuming a power law with coefficient m and exponent b for the sediment transport formula

$$s_s = m \, u^b \tag{12.2}$$

where u denotes flow velocity, and using the relation

$$Q = b \, h \, u \tag{12.3}$$

in which Q denotes discharge and h denotes water depth, one obtains

$$s_s = m \, Q^b B^{-b} H^{-b} \tag{12.4}$$

When a river flows into a sea or a lake with a constant water level, the water depth in the mouth of the river can be considered to be constant too because bed-level fluctuations will be negligible. Substitution of equation (12.4) into equation (12.1) leads to

$$h_{\text{mouth}} = \frac{m^{1/b} B^{(1-b)/b}}{S_m^{1/b}} \left[\frac{1}{T} \int_0^T Q^b \, dt \right]^{1/b} \tag{12.5}$$

This quantity is related to the bed level in the rest of the river. Elimination of h from equation (12.4) by using the Chézy equation gives

$$s_s = m B^{-b/3} C^{2b/3} Q^{b/3} i_b^{b/3} \tag{12.6}$$

in which C is the Chézy coefficient for hydraulic roughness and i_b is the longitudinal slope of the bed. Substitution with constant C into equation (12.1) yields

$$i_b = \frac{S_m^{3/b} B^{1-3/b}}{m^{3/b} C^2} \left[\frac{1}{T} \int_0^T Q^{b/3} \, dt \right]^{-3/b} \tag{12.7}$$

Equations (12.5) and (12.7) indicate that quantities representing the dynamic equilibrium of the bed are functions of an ath moment of the distribution of discharges over time, α_a, defined by

$$\alpha_a = \frac{1}{T} \int_0^T Q^a \, dt \tag{12.8}$$

The proper value of the exponent, a, depends on which quantity is being considered. The same applies, hence, to the value of the representative bed-forming discharge derived from the ath moment

$$Q_a = \sqrt[a]{\alpha_a} \tag{12.9}$$

This means that a single bed-forming discharge is different for different quantities of the dynamic equilibrium, and hence that "dominant discharge" is an inherently imprecise concept (NEDECO, 1959; Blench, 1969; Prins and de Vries, 1971; Jansen *et al.*, 1979). It is plausible, however, that the general form of the parameterization can also be used for relations with other morphological quantities, such as bank erosion.

Width adjustment

The hydraulic geometry of individual channels changes continuously due to changes in the supply and distribution of water and sediment at the upstream bifurcations. The distribution of discharges over the bifurcated channels can be calculated easily when the geometry of the channels is known, but the distribution of sediment transport is very difficult to assess. As this distribution depends on the geometry of the bifurcation, we try to find a correlation between width adjustment and upstream bifurcation geometry. The latter is represented by the deflection angle, i.e. the angle between a bifurcated channel and the channel upstream, see Figure 12.2, and by parameters expressing the degree of bifurcation asymmetry.

Channel abandonment

Channels are frequently abandoned as a result of chute cut-offs and channel avulsions. We try to correlate channel abandonment with the key parameters of the model for bend cut-offs by Klaassen and van Zanten (1989) as well as with the bifurcation geometry parameters used in the study on width adjustment.

Bend migration

Individual curved channels in the braided system can be seen as meandering rivers and nowadays several mathematical models for meander migration exist. Roughly,

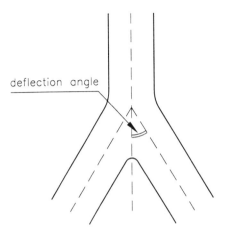

FIGURE 12.2 Definition of deflection angle

they can be subdivided into five categories. Listing them in the order of increasing complexity, these categories are:

1. Formulae in which bank migration rates are a function of local channel curvature (Hickin and Nanson, 1984).
2. Kinematic models in which bank migration rates are a function of local and upstream channel curvatures (Ferguson, 1983; Howard, 1983; Howard and Knutson, 1984).
3. Meander models based on the equations for water flow and bank migration in curved equiwidth channels (Ikeda et al., 1981).
4. Meander models based on the equations for water flow, sediment transport and bank migration in curved equiwidth channels (Johannesson and Parker, 1989; Crosato, 1990).
5. Bank migration models based on the equations for water flow, sediment transport and bank migration in channels with arbitrary geometries (Mosselman, 1992).

Klaassen and Masselink (1992) followed the approach of category (1). We attempt to improve this by simulating the migration of a few bends with a kinematic meander model like the ones in category (2). Bend migration in our model is not only a function of local and upstream channel curvatures, but also a function of the absolute value of the sine of the angle between channel centre-line and direction of maximum valley slope. The latter dependence is assumed to represent the downvalley migration due to cross-channel flow at high stages in analogy to the downstream propagation of a dredged trench.

The influence of the channel curvature at a certain upstream distance, d, depends on a characteristic length scale, L, according to

$$\text{weight factor} = \exp\left(-\frac{d}{L}\right) \tag{12.10}$$

The combined influence of a series of upstream curvatures is given by

$$\left(\frac{\partial n}{\partial t}\right)_a = \frac{\sum \left(\frac{\partial n}{\partial t}\right)_n \exp\left(-\frac{d}{L}\right)}{\sum \exp\left(-\frac{d}{L}\right)} \tag{12.11}$$

in which $(\partial n/\partial t)_n$ is the nominal bend migration rate, calculated from local channel curvature only, and $(\partial n/\partial t)_a$ is the adjusted bend migration rate.

Influence of tectonics

The Brahmaputra–Jamuna River lies in one of the most active tectonic zones of the world. The tectonic activity is related to the collision between the Indian and Eurasian plates (Molnar and Tapponier, 1977), the subsidence in this zone being the direct counterpart of uplift in the Himalayan belt. The activity has fractured the terrain into various geological compartments or blocks, separated by vertical fault planes. Differential uplift or subsidence and tilting of these compartments might affect the planform of the river, more or less like the planform of the Mississippi River which seems to respond to subtle uplift rates of the order of 3 mm per year (Watson et al. 1983).

We identify the faults by an analysis of lineaments on the satellite images and study correlations between these lines and the planform of the river. We use image processing techniques to enhance the contrast between various terrain units.

RESULTS

Relation between bank erosion and discharge

Figure 12.3 shows the relation between the total area of eroded land along 14 major bends and the fourth moment of the annual discharge record, a_4. A weak trend that the amount of bank erosion increases as the magnitude and the duration of the annual flood increases becomes visible. The trend is less clear when plotting lower moments of the discharge record with an exponent of, for instance, 1 or 2.

Inundated banks cannot be distinguished from eroded banks on the satellite images, so that higher stages produce spurious bank erosion on the images. The arrows in Figure 12.3 indicate the directions in which the points would move if a correction for this spurious erosion could be made. It must be noted, however, that bank-line errors due to spurious erosion are smaller than bank-line errors due to spurious accretion because the eroding banks are usually steep.

Width adjustment

We could not find a clear correlation between width adjustment and upstream bifurcation geometry. As an example, we show the relation between width adjustment and deflection angle in Figure 12.4. Annual width adjustment appears here as a random process with average zero and standard deviation equal to 0.46

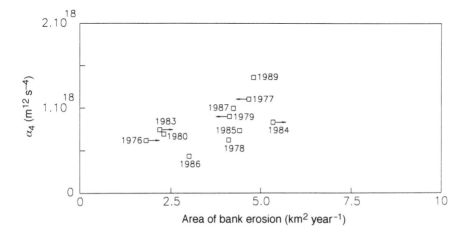

FIGURE 12.3 Relation between area of eroded land and fourth moment of flood hydrograph

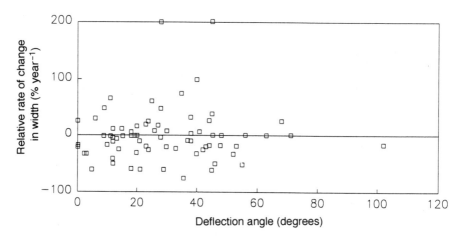

FIGURE 12.4 Relation between width adjustment and deflection angle

times the width. This result is nevertheless useful for the formulation of width adjustment in a probabilistic model.

Channel abandonment

Channel abandonment correlated better with the bifurcation geometry parameters than with the key parameters of Klaassen and van Zanten (1989). The relation between the frequency of channel abandonment and the deflection angle is shown in Figure 12.5. Comparing the clear trend in this figure with the randomness in Figure 12.4, we conclude that channels are abandoned through shallowing rather than narrowing. The slow bank accretion can be understood from the relatively small contribution of secondary currents and lateral turbulent diffusion to the transport and deposition of sediment. Hence most of the excess of sediment, which enters a channel over the full cross-section, will not reach the banks but only contribute to elevation of the bed.

Bend migration

Figure 12.6 shows, as an example, the results for the migration of a large bend at latitude 24° 4', close to the confluence with the Ganges River. The characteristic length scale in the relation between adjusted migration and upstream channel curvatures was 10 km. A sensitivity analysis revealed that computed planform changes are more sensitive to errors in the position of straight reaches than to errors in curved reaches.

Influence of tectonics

The major faults which we identified are presented in Figure 12.7. Details are reported by Hartmann et al. (1993). The general planform in Figure 12.7 shows very

Morphological Changes in a Large Braided Sand-Bed River 243

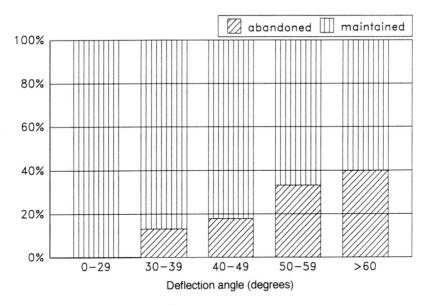

FIGURE 12.5 Relation between probability of channel abandonment and deflection angle

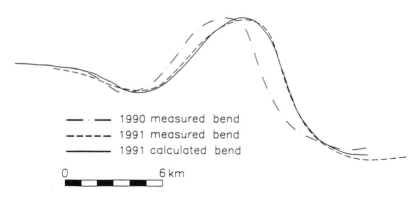

FIGURE 12.6 Comparison between observed and simulated channel migration for a large bend at latitude 24° 0'

few good correlations between river alignment and major faults, but channel patterns on a smaller scale within the braid belt suggest more strongly an influence of tectonics. First, the channels are sometimes remarkably straight in directions parallel to the faults. Second, changes in planform are sometimes difficult to explain from an autonomous motion of water and sediment. An example is shown in Figure 12.8, where the downstream translation of a bend at latitude 24° 10' seems to be inhibited between 1985 and 1987 by an apparent rectilinear obstacle complying with one of the faults which we identified in both the braid belt and the surrounding terrain. Field evidence of vertical terrain motions is needed to confirm these findings.

The usual opinion is that large-scale migration of the Brahmaputra–Jamuna River is primarily controlled by tectonics (e.g. Coleman, 1969) and that local pattern

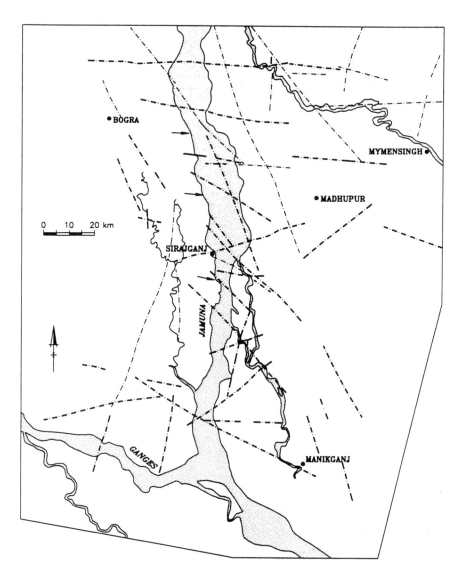

FIGURE 12.7 Major faults in the Bengal Basin as inferred from an analysis of lineaments from satellite images

changes within the braid belt can be understood from the dynamics of water and sediment alone. Our findings suggest that this view might be too simplistic because small-scale patterns seem to correlate better with fault lines than large-scale patterns. In addition, Thorne *et al.* (1993) propose for large-scale patterns that the general westward migration of the braid belt as a whole might also be explained as the development of a very large meander. This might be plausible as a residual effect of the planform changes of the individual channels. The overall curvature means that the sum of individual westward migrations will be larger than the sum of individual

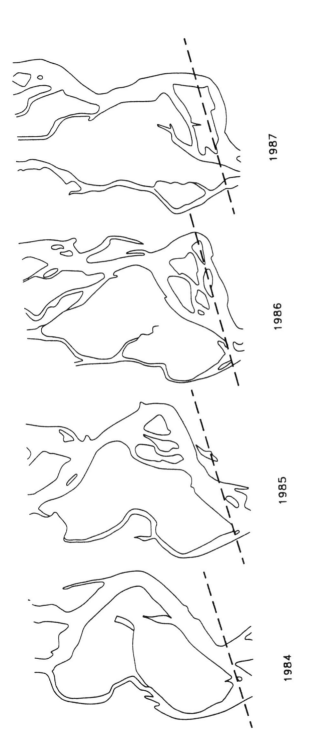

FIGURE 12.8 Inhibition of channel migration at a fault by tectonic block activity

eastward migrations, and also that on average the eastern channel below each bifurcation is more likely to be abandoned than the western channel. It thus seems that both large-scale patterns and small-scale patterns are influenced by tectonics as well as the dynamics of water and sediment.

CONCLUSIONS

We studied the planform changes in the braided Brahmaputra–Jamuna River by using satellite images, discharge records and cross-sectional data. The correlation between annual erosion and magnitude and duration of the annual flood was weak. Channel abandonment was clearly correlated with upstream bifurcation geometry, but channel width adjustment was not. From this we conclude that channels are abandoned through shallowing rather than narrowing. Bend migration of individual channels could be simulated with a simple meander model. There are indications that channel patterns within the braid belt are influenced by tectonics. In general we conclude that the phenomena are so convoluted that it is difficult to extract information on individual underlying mechanisms from observations only. It is necessary to bring descriptions of the underlying mechanisms together in a model to see whether their combined behaviour can reproduce the observed morphological changes. We have started the development of a probabilistic predictive model and its general concepts are described by Klaassen *et al.* (1993).

ACKNOWLEDGEMENTS

This study was carried out for FPCO (Flood Plan Coordination Organization, Bangladesh) with funding from KfW (Kreditanstalt für Wiederaufbau, Germany) and CFD (Caisse Française de Développement, France). It was initiated by Mr G. J. Klaassen, whom we thank for his support and encouragement. The support from Dr J. Rupke and Mr R. A. Hartmann (Alpine Geomorphology Research Group, University of Amsterdam), Professor J. Vandenberghe (Free University, Amsterdam) and Mr M. van der Wal (Delft Hydraulics) is gratefully acknowledged. We thank Dr T. Martin (FAP19) and Mr Y. Rauste (FINNMAP) for providing some of the digital satellite data, and Messrs H. H. S. Noorbergen and W. Verhoef (National Aerospace Laboratory NLR) for the processing of the satellite images.

REFERENCES

Blench, T. 1969. *Mobile-Bed Fluviology*. University of Alberta Press, Edmonton.
Bristow, C. S. 1987. Brahmaputra River: channel migration and deposition. In: *Recent Developments in Fluvial Sedimentology*, Society of Economic Paleontologists and Mineralogists, Spec. Publ. No. 39.
Coleman, J. M. 1969. Brahmaputra River: channel processes and sedimentation. *Sedimentary Geology*, **3**, 129–139.
Crosato, A. 1990. Simulation of meandering river processes. Communications on Hydraulic and Geotechnical Engineering, No. 90–3, Delft University of Technology.
Ferguson, R. I. 1983. Kinematic model of meander migration. In: Elliott, C. M. (ed.), *River Meandering, Proc. Conf. Rivers 1983*, New Orleans, ASCE, 1984, pp. 942–951.
Hartmann, R. A., Rupke, J. and Seijmonsbergen, A. C. 1993. The influence of neo-tectonic movements on the planform development of the Jamuna river, Bangladesh; A preliminary appraisal on the basis of Landsat MSS images. Alpine Geomorphology Research Group, Faculty of Environmental Sciences, University of Amsterdam (limited distribution).

Hickin, E. J. and Nanson, G. C. 1984. Lateral migration rates of river bends. *Journal of Hydraulic Engineering*, ASCE, **110**, 1557–1567.

Howard, A. D. 1983. Simulation model of meandering. In: Elliott, C. M. (ed.) *River Meandering, Proceedings Conference Rivers 1983*, New Orleans, ASCE, 1984, pp. 952–963.

Howard, A. D. and Knutson, T. R. 1984. Sufficient conditions for river meandering: a simulation approach.*Water Resources Research*, **20**, 1659–1667.

Ikeda, S., Parker, G. and Sawai, K. 1981. Bend theory of river meanders, Part 1, Linear development. *Journal of Fluid Mechanics*, **112**, 363–377.

Jansen, P. Ph., van Bendegom, L., van den Berg, J., de Vries, M. and Zanen, A. 1979. *Principles of River Engineering*. Pitman, London.

Johannesson, H. and Parker, G. 1989. Linear theory of river meanders. In: Ikeda, S. and Parker, G. (eds), *River Meandering, AGU, Water Resources Monograph 12*, pp. 181–213.

Klaassen, G. J. and Masselink, G. 1992. Planform changes of a braided river with fine sand as bed and bank material. *Proceedings Fifth International Symposium River Sedimentation*, Karlsruhe, pp. 459–471.

Klaassen, G. J. and van Zanten, B. H. J. 1989. On cutoff ratios of curved channels. *Proceedings of 23rd Congress IAHR*, 1989, Ottawa.

Klaassen, G. J. and Vermeer, K. 1988a. Channel characteristics of the braiding Jamuna River, Bangladesh. *Proceedings International Conference River Regime*, 1988, Wallingford.

Klaassen, G. J. and Vermeer, K. 1988b. Confluence scour in large braided rivers with fine bed material. *Proceedings International Conference Fluvial Hydraulics*, 1988, Budapest.

Klaassen, G. J., Mosselman, E. and Brühl, H. 1993. On the prediction of planform changes in braided sand-bed rivers. In: Wang, S. S. Y. (ed.), *Advances Hydro-Science and -Engineering*, The University of Mississippi, pp. 134–146.

Molnar, P. and Tapponier, P. 1977. The collision between India and Eurasia. *Scientific American*, **236**(4), 30–41.

Mosselman, E. 1992. Mathematical modelling of morphological processes in rivers with erodible cohesive banks. Communications on Hydraulic and Geotechnical Engineering, No. 92-3, Delft University of Technoloyy.

NEDECO (Netherlands Engineering Consultants) 1959. *River Studies and Recommendations on Improvement of Niger and Benue*. The Hague, the Netherlands.

Prins, A. and de Vries, M. 1971. On dominant discharge concepts for rivers. *Proc. 14th Congress IAHR*, Paris, Vol. 3, Paper C20.

Thorne, C. R., Russell, A. P. G. and Alam, M. K. 1993. Planform pattern and channel evolution of the Brahmaputra River, Bangladesh. In: Best, J. L. and Bristow, C. S. (eds), *Braided Rivers, Geological Society Spec. Publ.* No. 75, pp. 257–276.

Watson, C. C., Schumm, S. A. and Harvey, M. D. 1983. Neotectonic effects on river pattern. In: Elliott, C. M. (ed.), *River Meandering, Proc. Conf. Rivers 1983*, New Orleans, ASCE, 1984, pp. 55–66.

Index

Accretion, 179
Aggradation facies, 179
Alluvial deposits, 195
Alluvial terrace, 188
Almanzora River, south-east Spain, 171
Arbucies River, south-east Spain, 93–102
    channel cross-section, 96
    drainage basin, 95
    flash flood magnitude index, 96
    flow duration variability index, 96
    flow-duration curve, 95, 97
    mean annual base flow, 96
Armour layer, 97, 100, 156, 160
Avalanche face, 186, 188

Bangladesh Water Development Board, 236
Bank,
    accretion, 242
    breaches, 136
    erodiblity, 132
    erosion, 124, 174, 177, 189, 204–6, 239, 241
        and discharge, 237–9, 241
        rate of, 183, 217, 235
    failure by liquifaction, 183
    migration, 240
    protection, 203, 206, 222
        structures, 193, 235
    stability, 189
Bar,
    growth, 124
    head, 124
    supraplatform, 183
Bars, 40, 85–7, 183, 186, 188
    marginal, 186
    medial, 113, 124
    point, 56, 79, 85
    re-attachment, 55, 81, 85

Base-level, 191
    change, 190
Basin sedimentation, 221
Bed,
    aggradation, 128
    deformation, 161–2
    elevation, 161–2, 164, 166,
        and discharge, 163
    geometry, 63
    load, 33, 35
    material armouring, 40
    material, specific gravity of, 40
    microtopography, 44, 109
    particle clusters, 55–6, 61, 66, 69, 71, 75, 87
        complex, 75
        diamond arrangement, 58, 71–3, 86
        multiple obstacle 75–6
        obstacles, 73
        orientations, 69
        rhomboidal lattice structure, 74
    particle shape, 37, 45, 47, 49, 52
        influence on entrainment shear stress, 49
        influence on transport, 47–51
        rounding, 64
    particle size, 39, 64, 115
        determination, 64
        distribution skewness, 46
        fourier coefficients, 64
        fourier spectrum, 64
        gamma distributions, 37, 50
    particles, degree of projection of, 61, 63
        entrainment probability and transport of, 38, 44, 47
        hiding effects of, 37
        orientation of, 64
        roughness of, 44
        step length of, 51

Bed (*cont.*)
  structural locking of, 44
  transport distances of, 37–8, 40–1, 44–7, 49–52
    influence of particle weight, 45–7, 52
    influence of shape, 37, 45, 47, 49, 52
    travel pathways, 38
  roughness, 128
  scour, 156, 177
  topography, 110, 112
    and particle characteristics, 38
    lobate cross-channel ribs, 56
    point bars, 57
    pools, 47, 52, 56
    riffles, 56
    step-pool, 37, 39–40, 52–3, 56, 66, 83–5
    step reaches, 47
    transverse ribs, 56, 66
Bed-level fluctuations, 238
Bed-load transport rate, 34
  measured over dunes, 30–2
Bed-material load, 30, 34
Bed-material movement, 173
Bedform, 66, 115, 188, 190
  adjustments, 79, 174
  geometry and migration rates, 21
  height, 27
  hysteresis, 27
  migration, 100–1
  roughness, 55–6
  shape factor, 27
  transverse, 73
Bedforms, 19, 73, 99
  and initiation of particle motion, 99
  dunes, 19
  flow-transverse, 19
  longitudinal, 73
  megaripples, 19
  ripples, 19
  waves, 19
Bedload, 224
  exchange and budget, 44
  movement, 94, 174, 179
  sediment, 40
  source areas, 44
  transport, 47, 93–4, 96–8, 101, 106, 178, 183
  yield, 98–100
Bend,
  curvature, 57
  migration, 235, 239–40, 242, 246
    kinematic models of, 240

Benipila *Rambla*, south-east Spain, 171–4, 177, 180, 183, 188, 190–1
Berms, cobble, 55, 57, 80–2
Betic Range, south-east Spain, 171, 173
Brahmaputra-Jamuna River, eastern India and Bangladesh, 235–46
Braid belt, 237, 243
  migration, 244
Braided river, 58, 126, 169, 173, 178, 181, 190, 212–13, 216, 235

Caesium-137, 146–7, 150
Canalization, 189–91,
Carpathians, 210–11, 213
Carpathian rivers, 230
Catastrophic flood events, 190
Centrifugal force, 55, 71
Channel,
  abandonment, 136, 151, 235–6, 239, 242, 246
    and deflection angle, 243
    frequency, 242
  adjustment,
    pattern and process of, 105–30
  adjustments (to floods), 179–80
  aggradation, 131, 181, 213
  avulsion, 239
  banks,
    basal erosion of, 188
    concave and convex, 205
  bifurcation, 246
    asymmetry, 239
    deflection angle, 239
    geometry, 239, 241–2, 246
  change, 127, 195, 209, 212–13, 239,
  conveyance, 237
  curvature, 66, 85, 240
  degradation, 181
  deposition, 180, 184
  downcutting, 230
  erosion, 180, 184, 230
  gradient, 66, 139
  migration, 131, 242
    and faulting, 245
  migration, downvalley, 240
  pattern, 179, 209, 218–19, 221, 235, 237
  regulation, 217, 226
  scour and fill, 155, 162
  shifting, 200
  sinuosity, 143, 183, 193
  widening, 230, 235, 239–41
    and deflection angle, 242
  width/depth ratio, 86

Channel/floodplain interface, 133, 135
Channelization, 209–29
Check dams, 51, 61
Chezy equation, 174, 238
Chute cutoff, 239
Clasts, open-bed, 69, 70–1
Cohesive material, 132
Columbia River, USA, 30, 32
Computational grid, 110
Concentration gradient, 136
Confluence, 126
  angle, 126
  bifurcations, 235
  migration, 235
Continuity of mass equation, 133
Convective transport, 109, 134
Crevasse,
  channels, 229
  splays, 190
Cross-stratification, 186
Current speed and direction, 21
Cutoff, 239

Dams, 191, 209, 211–12, 230
Darcy-Weisbach friction factor, 109, 115
Debris,
  cone, 183
  end-lobe, 87
  flow, 40, 58, 60, 79, 86–7, 189
Deforestation, 209
Degradation potential, 128
Deposition, 242
  on floodplains, 136, 143, 145, 150–1
Depositional reach, 58
Diagonal rib, 78
Diffusive transport, 109, 134
Diffusivity, vertical, 135
Digital terrain model (DTM) 105, 113, 115, 128
Discharge,
  bankfull, 93, 94, 166, 216, 222, 224, 237
    frequency and channel adjustments to, 97–101
  channel-forming, 216, 224, 238–9
  dominant, 93–4, 101, 239
  effective, 93–102, 94, 100
  effectiveness, 178
  thresholds, 181
Drainage network, 198
Dunajec River, southern Poland, 213, 216
Dunes in Fraser River Estuary, BC Canada, 19–35
  and sediment transport, 19–36
  form/process hysteresis, 30
  height, 21, 23–5, 29, 34
  length, 21, 23–5, 29, 34
  migration, 20, 23–5, 29, 30, 33–4
  steepness, 23–5, 29–30
Dynamic equilibrium, 169, 170, 173, 188–9, 191

Echosounding profiles, 23–4
Eddy,
  diffusivity, 134
  viscosity, 108–11, 126
Electromagnetic current meter (ECM), 115–16, 121
End lobes, debris, 87
Energy slope, 56
Entrainment, 127
  function, 128
  threshold, 128
Ephemeral channels, 169–70, 173, 179, 190
Episodic fluvial sedimentation, 170
Epsilon cross-stratification, 188
Erosional capacity, 205
Erosion pins, 174
Event frequency, 170
Extreme events, 181

Flood,
  control structures, 226–7
  deposits, 170
  hydraulics, 71, 151
  hydrograph, 44
  hydrology, 58
  peak, 175, 190
Flooding, 39, 49, 187
Floodplains, 136, 139, 150–1, 179, 183, 217, 220
  alluviation, 230
  backwater areas, 136
  degradation, 135
  deposition, 151
  development, 179
  geomorphology, 138, 143, 210–11, 222, 226, 229
  inter-embankment zone, 213, 226, 229, 230
  inundation, 131, 139, 142, 227
  management, 235
  sediment, 141
  sedimentation patterns, 131
  sedimentation rates, 131
Floods, 169
  frequency and magnitude of, 213

Floodwater velocities, 174
Flow,
  bifurcation, 124–5
  convergence, 109
  divergence, 109, 124
  expansion zone, 55
  in rill channels, 9, 10
  in rills, 5, 13
  overland concentrated, 5
  patterns, 125
  resistance, 44
  separation, 30, 32, 34
Flow-duration curve, Arbucies River, Spain, 95, 97
Form drag, 61
Fraser River at Marguerite, BC Canada, 155–66
Fraser River Estuary, BC Canada, 19–35
Friction angle of sediment, 30, 32, 34
Froude number, 74

Gamma,
  distributions, 53
  spectrometry, 150
Ganges River, 242
Geomorphic processes, magnitude and frequency of, 94
Geomorphological effectiveness (of floods), 177–80, 183
Gradient, valley, 139
Grain-size, 22
  effective, 131, 140, 142–5, 148, 150–1
  floodplain sediments, 136, 143
  ultimate, 131, 143–5, 148, 150–1
Grain-size distributions, 40, 145
  ultimate, 142
Gravel bar, 73, 183
Gravel bar morphology, 188
Gravel-bed rivers, 157
Gravel sheets, 183
Groynes, 203, 205–6, 221, 226
Groyne system, 217, 229
Gullies, 188
Gumbel frequency distribution, 175

Haut Glacier d'Arolla (south-west Switzerland), 105–30
Helical flow, 110
Human impact on rivers, 206, 209, 211, 213–14, 226, 230
Hydraulic geometry, 57, 86, 155–66
  at-a-station, 155–6, 160
  downstream, 155

Hydraulic,
  jumps, 74
  relations and threshold scour discharge, 164–6
Hydrograph recession, 186
Hydrologic regime, 106, 198
Hydrologic response, 226
Hydrologic thresholds, 170
Hydrostatic pressure, 74
Hyperconcentrated flow, 86
Hysteresis effects, 26–7, 29, 162

Imbrication, 47, 52, 55, 61, 66–8, 75, 81
  angle, 61
Incipient motion of bed material, 69, 166
Incised meanders, 222
Incision, 210
Irrigation, 191

Jucar River, south-east Spain, 171

Kotlaine River, Bavaria, 37–8

Lainbach River, Bavaria, 37–53, 58
Landuse in river catchments, 209
Laser diffraction particle-size analyser, 142–3
Lateral accretion, 179
  deposits, 131
Lateral migration, 217
  rates, 230
Law-of-the-wall, 109, 128
Levees, 136, 143, 200, 203–4, 213, 217, 222, 224, 226–7, 229
  arch-shaped, 81
  artificial, 198
Lineaments on satellite images, 241
Little Ice Age, 213, 216, 222
Load,
  bed-material, 19–20
  continuous suspension, 19
  intermittent suspension, 19
  wash, 19, 22
Lobate rib, 78
Log Gumbel frequency distribution, 175
Log jams, 55, 57–8, 60, 66, 87
Log-Pearson Type III frequency distribution, 175
Log-transform bias, 100
Long profile of river bed, 40, 56, 237
Low-sinuosity channel, 169, 185

Magnitude and duration of effective discharge, 102
Magnitude and frequency of torrential events, 170
Magnitude-frequency analysis of discharge, 178
Manning equation, 133–4
Manning roughness coefficient, 109, 111, 114–15, 127, 133–4
Mass,
 balance equation, 134
 failure, 183
 movement, 40, 229–30
Meander, 200, 203, 205–6, 216–17, 226
 belt, 136, 139, 143, 193, 206
 channels, 136, 169, 179, 188, 203, 212–13, 216–17, 224, 239
 cutoffs, 212–13, 215, 226, 229
 migration, 239
 models, kinematic, 240
 radius, 216
 wavelength, 216
Meandering valleys, 58
Megaclusters of bed material, 57, 79
Megaripples, 183
Microprofiling of bed topography, 60–3
Microtopography, 115
Minimum rate of energy dissipation, theory of, 170
Mining, 209
Mixing-length, 109
Model optimization, 124
Model,
 two equation k-ε, 109, 110–11
 zero-equation, 108–10
Momentum and inertial effects, 126
Momentum equation, 110
Momentum exchange in floodplain flows, 132
Mountain valley form and river bed arrangement, 55–88

National Rivers Authority, UK, 145
Navier-Stokes equations, 107
Nogalte 'Rambla', south-east Spain, 171–4, 178–83, 190–91

Overbank,
 deposits, 228
 flooding, 132, 136
 flow, 174, 179
 sedimentation on floodplains, 131–51
Oxbow lakes, 227, 229

Paleo-river channels, 204–5
Particle fall velocity, 135
Particle size, 142, 151
 analysis, 143
 distributions, 148–50
 transported material, 99
Photo-sieving of bed material, 63–4, 65
Piedmont Vistula, southern Poland, 211–30
Planform changes, 244
Po River basin, Northern Italy, 194
Po River, Mantova Province, northern Italy, 193–206
 evolution of, 196–205.
 migration of, 198
Point bars, 56, 85, 179
Polish Hydrological Survey, 212, 231
Pollution of rivers, 211
Probabilistic approaches to modelling, 128
Pump sample, 22

Quadratic friction law, 128

Raba River, southern Poland, 217
Ramblas, 171, 173
Rating curve, 164, 222
Recurrence interval,
 of effective discharge, 178
 of floods, 175–6, 190
Relative roughness, 63, 156
Resistance to flow, 166, 186
Reynolds number, 108
Reynolds stresses, 109
Riffles, 125, 186
Rills, 58
 depth, 9
 feeders, 5
 network, 5
 system, 5
 thalweg, 7, 13
 width, 9
Ripples, 188
River Culm, Devon, UK, 131–51
River Severn, UK, 145
River,
 alignment and faulting, 243
 bed topography, 41
 diversion, 197, 206
 network, 213
 terrace, 188
 training, 235
River-bed morphology, 57
Roughness,
 assemblages, 87

Roughness (*cont.*)
  bed, 86, 105
  bedform, 86, 88
  channel, 57, 61, 106
  distribution of, 81
  elements, 49, 61, 74, 81, 87, 116
  form and system, 56–7, 66–79, 80–7
  grain, 56
  hydraulic, 238
  spatial, 64
  three-dimensional, 63
  unit, 86
  variability and flow, 67–71
Runoff, 5
  surface, 2, 5

Salt wedge, 20
San River, southern Poland, 213
Sand-bed rivers, 157
Sand waves, straight crested, 183
Satellite images, 235–46
Schmiedlaine River, Bavaria, 37–53
Scour, 131, 164, 189
  chains, 173
  depth, 173, 235
Secondary circulation, 86, 186
Secondary currents, 242
Secondary flow, 55, 110
Sediment,
  aggregates, 13–14, 16
  concentration, 22, 135–6, 143, 170
    gradient, 143
  density, 29, 150
  dispersed, 14, 16, 142
  load, 212, 230
  particle-size distribution, primary, 7
  sampler,
    depth-integrated, 174
    Helley-Smith, 21–2, 30, 32–4, 96
      trapping efficiency, 96
    pump 23
    USP-61 point-integrating river, 23
  sequence, upward-coarsening, 183
  size,
    distribution, 1
      dispersed sediment, 8–9, 11
      effective, 2
      eroded sediment, 2, 9–11, 13
      suspended sediment, 12
      transported sediment, 2
      undispersed sediment, 6
    of sediment eroded from agricultural soil, 1–17
    primary, 9
  textural classification, 15
  texture, 14, 16
  transport, 20, 29, 32, 169, 178–9, 209–10, 230, 238, 240, 242
    estimates based on dune migration, 27–30
    of coarse bedload, 40
    rate, 27, 226
  undispersed, 6, 14, 16
Sedimentation, 204–5, 214, 226–7
  overbank, 213, 222, 224, 226, 229–30
  rates of, 132, 226
  rate, 150, 209, 212, 230
  traps, 143
Segura River, Murcia, Spain, 169–91
Shear stress, 44, 47, 49, 127–8, 183
  critical, 99
  from velocity profiles, 109
shear waves, 73–4, 86
Shields entrainment function, 99
Slope, water-surface, 26
Slumps, bank, 58
Soils, agricultural, 2
Sola River, southern Poland, 217
Stage/discharge relation, 134
Step pools, 83–5
Step-pool systems, 84–5
Stochastic approaches to modelling, 128
Stoke's velocity law, 108
Stream power, 51, 170
  threshold, 178
STREMR model, 109–11, 113, 116, 120
Strickler equation, 115
Subsidence, 198
Surface waves, rhomboidal, 74
Suspended load/flood discharge hysteresis, 179
Suspended sediment, 15, 143, 174
  concentration, 21, 30, 134, 141–2, 145, 150
  distribution across floodplain, 133
  load, 32, 33, 132, 140, 142, 171, 177, 179, 181, 211, 224, 229
  rating curve, 166
  samples, 22
  transport, 227

Tacheometric survey, 113
Tausendfussler micro-profiler, 44, 61–3
Tectonic control of rivers, 193, 198, 206, 235–6, 242–3, 246
Terraces, river, 182, 190
Terrestrial analytical photogrammetry, 112–13

Threshold scour discharge, 156, 162–6
Tidal height, 26–7
Tides, 20
Tordera River, south-west Spain, 94
Torrential flow and channel morphology, 169–91
Torrential flows in ephemeral channels, 174
Torrential runoff, 179
Torrential streams, 170
Tracers, 40
  concrete, 41
  magnetic, 37–8, 41, 47, 51–3
  plastic, 37, 41
  radio, 51
  shape, 41–3
  starting positions and deposition sites, 44
Transport,
  bed material, 19, 21, 32,
  suspended-sediment, 20
Transverse bar, 188
Transverse rib, 75, 77–8, 84
Transverse steps, 77
Turbulence, 106, 108, 110, 143
Turbulent diffusion, 132, 242
Turbulent fluid flow, computational models of, 106

Urbanization, 209, 230

Vegetation, woody, 189
Velocity, 7, 13, 27, 30, 121
  cross-sectional, mean, 26
  gradient, 164
  patterns, 125
  vectors, 123
Velocity/discharge gradient, 164–5
Vistula River, southern Poland, 209–31
Vortex shedding, 109

Washload, 34
Water elutriation system, 140
Water-surface,
  function, 133–4, 139–40
  elevation, 156, 161, 164
  slope, 159, 164
  topography, 112, 127
Water Survey of Canada, 155, 157
Waves, standing, 84
Wisloka River, southern Poland, 217, 221
Wolman method, 64

Yampa River basin, USA, 94